American Book Company
The Standards Experts

MASTERING THE GRADE 7 COMMON CORE

IN

MATHEMATICS

ERICA DAY
COLLEEN PINTOZZI
AMANDA CAHOY
MARY ANNE CRISCI
MARY REAGAN

REVIEWED BY:
ANGELA REDMAN

AMERICAN BOOK COMPANY

P. O. BOX 2638

WOODSTOCK, GEORGIA 30188-1383

TOLL FREE 1 (888) 264-5877   TOLL FREE FAX 1 (866) 827-3240

WEB SITE: www.americanbookcompany.com

Acknowledgements

In preparing this book, we would like to acknowledge Melvin Carter and Breanna Bloomfield for their contributions in editing, Camille Woodhouse for her contributions in formatting, and Mary Stoddard for her contributions developing graphics for this book. We would also like to give a special thanks to Jessica DeBord for creating the teaching activities located in the answer key and our many students whose needs and questions inspired us to write this text.

Chart of Standards

Standard	Chapter Number
7.RP.1	6
7.RP.2	7
7.RP.3	5, 6
7.NS.1	1, 2, 3
7.NS.2	1, 2, 4
7.NS.3	3, 4
7.EE.1	9
7.EE.2	8
7.EE.3	9
7.EE.4	8, 9
7.G.1	11
7.G.2	11
7.G.3	13
7.G.4	12
7.G.5	10
7.G.6	11, 13
7.SP.1	14
7.SP.2	14
7.SP.3	14
7.SP.4	14
7.SP.5	15
7.SP.6	15
7.SP.7	15
7.SP.8	15

Contents

Acknowledgements   ii

Preface   x

1 Integers   **1**
   1.1   Integers (DOK 1)   1
   1.2   Absolute Value (DOK 1)   1
   1.3   Opposite Numbers (DOK 1)   2
   1.4   Word Problems Using Opposites (DOK 2)   3
   1.5   Adding Integers (DOK 1)   4
   1.6   Rules for Adding Integers with the Same Signs (DOK 1)   5
   1.7   Rules for Adding Integers with Opposite Signs (DOK 1)   6
   1.8   Subtracting Integers (DOK 1)   7
   1.9   Multiplying Integers (DOK 1)   8
   1.10   Dividing Integers (DOK 1)   8
   1.11   Rules for Multiplying and Dividing Integers (DOK 1)   8
   1.12   Going Deeper into Integers (DOK 3)   9
         Chapter 1 Review   10
         Chapter 1 Test   11

2 Rational Numbers   **13**
   2.1   Rational Numbers (DOK 1)   13
   2.2   Comparing Fractions (DOK 2)   14
   2.3   Comparing Decimals (DOK 2)   15
   2.4   Changing Fractions to Decimals (DOK 1)   16
   2.5   Changing Decimals to Fractions (DOK 1)   17
   2.6   Graphing Rational Numbers on a Number Line (DOK 2)   19
   2.7   Comparing Rational Numbers (DOK 2)   21
   2.8   Going Deeper into Rational Numbers (DOK 3)   22
         Chapter 2 Review   23
         Chapter 2 Test   24

3 Adding and Subtracting Rational Numbers   **25**
   3.1   Simplifying Improper Fractions (DOK 1)   25
   3.2   Adding Fractions (DOK 1)   26

Contents

3.3   Subtracting Mixed Numbers from Whole Numbers (DOK 1)   27

3.4   Subtracting Mixed Numbers with Borrowing (DOK 2)   28

3.5   Adding Decimals (DOK 1)   29

3.6   Subtracting Decimals (DOK 1)   30

3.7   Representing Addition and Subtraction on a Number Line (DOK 2)   31

3.8   Real World Addition and Subtraction Problems (DOK 2)   32

3.9   Going Deeper into Adding and Subtracting Rational Numbers
        (DOK 3)   33

        Chapter 3 Review   34

        Chapter 3 Test   35

4   Multiplying and Dividing Rational Numbers   **37**

4.1   Multiplying Fractions (DOK 1)   37

4.2   Dividing Fractions (DOK 1)   38

4.3   Multiplication of Decimals (DOK 1)   39

4.4   More Multiplying Decimals (DOK 1)   40

4.5   Division of Decimals By Whole Numbers (DOK 1)   41

4.6   Division of Decimals by Decimals (DOK 1)   42

4.7   Real World Multiplication and Division Problems (DOK 2)   43

4.8   Going Deeper into Multiplying and Dividing Rational Numbers
        (DOK 3)   44

        Chapter 4 Review   45

        Chapter 4 Test   46

5   Percents   **47**

5.1   Changing Percents to Decimals and Decimals to Percents (DOK 1)   47

5.2   Changing Percents to Fractions and Fractions to Percents (DOK 1)   48

5.3   Percent Word Problems (DOK 2)   49

5.4   Finding the Percent of the Total (DOK 2)   50

5.5   Percent Increase or Decrease (DOK 2)   51

5.6   Tips and Commissions (DOK 2)   53

5.7   Finding the Amount of Discount (DOK 2)   54

5.8   Finding the Discounted Sale Price (DOK 2)   55

5.9   Markups (DOK 2)   56

5.10   Sales Tax (DOK 2)   57

5.11   Understanding Simple Interest (DOK 2)   58

5.12   Percent Error (DOK 2)   59

5.13   Going Deeper into Percents (DOK 3)   60

Chapter 5 Review   61

Chapter 5 Test   62

6 Rates, Ratios, and Proportions   **65**

6.1   Rate (DOK 2)   65

6.2   More Rates (DOK 2)   66

6.3   Ratio Problems (DOK 2)   67

6.4   Expressing Ratios as Percents and Percents as Ratios (DOK 2, 3)   68

6.5   Solving Proportions (DOK 1)   71

6.6   Ratio and Proportion Word Problems (DOK 2, 3)   72

6.7   Mixture Problems (DOK 3)   74

Chapter 6 Review   75

Chapter 6 Test   76

7 Proportional Relationships   **78**

7.1   Proportional Relationships (DOK 1)   78

7.2   Finding Relationships in Tables (DOK 2)   79

7.3   Finding Relationships in Graphs (DOK 1)   80

7.4   Finding Relationships in Equations (DOK 1)   81

7.5   Graphing Linear Data (DOK 3)   82

7.6   Applying Proportional Relationships (DOK 3)   85

7.7   Comparing Proportional Relationships (DOK 2)   88

Chapter 7 Review   90

Chapter 7 Test   92

8 Introduction to Algebra   **94**

8.1   Algebra Vocabulary (DOK 1)   94

8.2   Substituting Numbers for Variables (DOK 1, 2)   95

8.3   Understanding Algebra Word Problems (DOK 1, 2)   96

8.4   Setting Up Algebra Word Problems (DOK 2)   100

8.5   Changing Algebra Word Problems to Algebraic Equations (DOK 2)   101

8.6   Going Deeper into the Introduction to Algebra (DOK 3)   102

Chapter 8 Review   103

Chapter 8 Test   104

9 Equations and Inequalities   **105**

9.1   Two-Step Equations (DOK 2)   105

9.2   Two-Step Equations with Rational Numbers (DOK 2)   106

Contents

9.3   Combining Like Terms (DOK 1)   108

9.4   Removing Parentheses (DOK 2)   108

9.5   Solving Two-Step Algebra Word Problems (DOK 2)   110

9.6   Graphing Inequalities (DOK 2)   111

9.7   Solving Inequalities by Addition and Subtraction (DOK 2)   113

9.8   Solving Inequalities by Multiplication and Division (DOK 2)   114

9.9   Two-Step Inequalities (DOK 2)   115

9.10   Solving Inequality Word Problems (DOK 2)   117

9.11   Going Deeper into Equations (DOK 3)   118

      Chapter 9 Review   119

      Chapter 9 Test   121

10 Angles   **123**

10.1   Angles (DOK 1)   123

10.2   Types of Angles (DOK 1)   124

10.3   Adjacent Angles (DOK 1)   125

10.4   Vertical Angles (DOK 2)   126

10.5   Complementary and Supplementary Angles (DOK 2)   127

10.6   Finding Angles in Figures (DOK 2)   128

      Chapter 10 Review   129

      Chapter 10 Test   130

11 Plane Geometry   **131**

11.1   Types of Triangles (DOK 1)   131

11.2   Types of Quadrilaterals (DOK 1)   132

11.3   Quadrilaterals and Their Properties (DOK 1)   133

11.4   Types of Polygons (DOK 1)   134

11.5   Drawing Shapes (DOK 2)   135

11.6   Area of Squares and Rectangles (DOK 1)   142

11.7   Area of Triangles (DOK 2)   143

11.8   Area of Trapezoids and Parallelograms (DOK 2)   144

11.9   Area of Composite Figures (DOK 2)   145

11.10   Area Word Problems (DOK 2)   147

11.11   Scale Drawings (DOK 2)   148

11.12   Solving Word Problems Using Scale Drawings (DOK 2)   149

11.13   Solving Scale Drawings Using Figures (DOK 2)   150

11.14   Going Deeper into Plane Geometry (DOK 3)   152

Chapter 11 Review 153

Chapter 11 Test 155

12 Circles **157**

12.1 Parts of a Circle (DOK 1) 157

12.2 Circumference (DOK 1) 158

12.3 Area of a Circle (DOK 1) 160

12.4 Relating Circumference and Area (DOK 2) 162

12.5 Circle Word Problems (DOK 2) 163

12.6 Going Deeper into Circles (DOK 3) 164

Chapter 12 Review 165

Chapter 12 Test 166

13 Solid Geometry **168**

13.1 Solid Figures (DOK 1) 168

13.2 Cross Sections (DOK 1) 169

13.3 Understanding Volume (DOK 1) 171

13.4 Volume of Cubes and Rectangular Prisms (DOK 2) ... 171

13.5 Volume of Other Prisms (DOK 2, 3) 174

13.6 Volume of Pyramids (DOK 2) 175

13.7 Surface Area of Cubes and Rectangular Prisms (DOK 2) ... 176

13.8 Surface Area of Other Prisms (DOK 2) 178

13.9 Surface Area of Pyramids (DOK 2) 179

13.10 Surface Area of Composite Figures (DOK 3) 180

13.11 Volume and Surface Area Word Problems (DOK 3) 181

13.12 Going Deeper into Solid Geometry (DOK 3) 182

Chapter 13 Review 183

Chapter 13 Test 185

14 Statistics **187**

14.1 Range (DOK 1) 187

14.2 Mean (DOK 1) 188

14.3 Finding Data Missing From the Mean (DOK 2) 189

14.4 Median (DOK 1) 190

14.5 Mode (DOK 1) 191

14.6 Applying Measures of Central Tendency (DOK 3) 192

14.7 Comparing Two Sets of Data (DOK 2, 3) 193

14.8 Simple Random Sampling (DOK 2) 195

Contents

14.9   Representative Sampling (DOK 3)   196

Chapter 14 Review   197

Chapter 14 Test   198

15 Probability   **199**

15.1   Probability (DOK 2)   199

15.2   Independent and Dependent Events (DOK 2)   201

15.3   More Probability (DOK 2)   203

15.4   Simulations (DOK 3)   204

15.5   Tree Diagrams (DOK 2)   206

15.6   Probability Models (DOK 3)   208

15.7   Predictions Using Probabilities (DOK 3)   210

15.8   Equally Likely Versus Equally Probable (DOK 2)   212

Chapter 15 Review   213

Chapter 15 Test   215

16 How to Answer Constructed-Response Questions   **217**

16.1   Writing Short Answers   217

16.2   Writing Open-Ended Answers   218

Chapter 16 Review   220

Index   **221**

Preface

Mastering the Grade 7 Common Core in Mathematics will help you review and learn important concepts and skills related to 7th grade mathematics. **The materials in this book are based on the common core standards in mathematics coordinated by the National Governors Association Center for best Practices and the Council of Chief State School Officers. The complete list of standards is located at the beginning of the Answer Key. Each chapter is referenced to the standards.**

This book contains chapters that teach the concepts and skills for 7th Grade. Answers to the tests and exercises are in a separate manual.

Mastering the Grade 7 Common Core in Mathematics includes Depth of Knowledge levels for four content areas based on Norman Webb's Model of interpreting and assigning depth of knowledge levels to both objectives within standards and assessment items for alignment analysis.

The four levels of Depth of Knowledge:

Level 1: Recall and Reproduction (DOK 1)

Questions at this level (DOK 1) include the recall of information such as facts, definitions, or simple procedures. As well as, performing a simple algorithm, or carrying out a one-step, well-defined, and straight forward procedure. A few example DOK level 1 questions are listed below.

· State the associative property of multiplication

· Identify the divisor in a problem

· Measure the perimeter of a figure

· Calculate $4.3 + 8.5$

Level 2: Skills and Concepts/Basic Reasoning (DOK 2)

DOK level 2 questions involve some mental processing beyond a habitual response. It requires students to make some decisions as to how to approach the problem. As well as being able to classify, organize, estimate, make observations, collect, display, and compare data. A few example DOK 2 level questions are listed below.

· Interpret the bar graph to answer questions about a given population

· Classify different types of polygons based upon their characteristics

· Compare two sets of data using their measures of central tendencies

· Extend an algebraic pattern

Level 3: Strategic Thinking/Complex Reasoning (DOK 3)

This level (DOK 3) includes problems that require reasoning, planning, using evidence, and higher levels of thinking beyond what was required in DOK levels one and two. This level requires students to explain their thinking, and cognitive demands are more complex and abstract. DOK 3 demands that students use reasoning skills to draw conclusions from observations and make conjectures. Some examples of level 3 DOK questions are listed below.

· Explain how you can determine if two triangles are similar

· Formulate an expression to determine the next few terms in a pattern

· Construct a survey and analyze the results to determine the most popular movie genre

Level 4: Extended Thinking/Reasoning (DOK 4)

DOK level 4 questions include things such as: complex reasoning, planning, and developing. Student thinking will most likely take place over an extended period of time, and that will include taking into consideration a number of variables. Students should be required to make several connections and relate ideas within the content area or among other content areas. By selecting one approach among many alternatives on how a situation should be solved. At this level students will be expected to design and conduct their own experiments, make connections between findings and relate them to concepts and phenomena together. A few example problems are presented below:

· Explore real world phenomena of Cartesian plans and create a report to present your findings

· Connect your knowledge of integers to the plate tectonics of Earth

· Analyze common game pieces (i.e. dice, spinners, etc.) to determine their fairness based upon what you know about probability of events by designing and carrying out your own experiment

ABOUT THE AUTHORS

Erica Day has a Bachelor of Science Degree in Mathematics and working on a Master of Science Degree in Mathematics. She graduated with high honors from Kennesaw State University in Kennesaw, Georgia. She has also tutored all levels of mathematics, ranging from high school algebra and geometry to university-level statistics, calculus, and linear algebra.

Colleen Pintozzi has taught mathematics at the middle school, junior high, senior high, and adult level for 22 years. She holds a B.S. degree from Wright State University in Dayton, Ohio and has done graduate work at Wright State University, Duke University, and the University of North Carolina at Chapel Hill. She is the author of many mathematics books including such best-sellers as *Basics Made Easy: Mathematics Review, Passing the New Alabama Graduation Exam in Mathematics, Passing the Louisiana LEAP 21 GEE, Passing the Indiana ISTEP+ GQE in Mathematics, Passing the Minnesota Basic Standards Test in Mathematics,* and *Passing the Nevada High School Proficiency Exam in Mathematics.*

Amanda Cahoy is pursuing a Bachelor of Arts degree in mathematics and chemistry at Dartmouth College. She has a varied background in mathematics, with endeavors ranging from participating in Georgia's highly selective Governor's Honors Program as a math "major" and organizing a math tournament for middle school teams. She was also a valued member of her high school's math team. Tutoring in her spare time, she enjoys helping other students succeed and has experience with levels ranging from elementary school to college calculus. She joined the American Book Company as a mathematics intern during the summer of 2011 before returning to Dartmouth College in the fall.

Mary Anne Crisci has a Bachelor of Science Degree in Mathematics. She recently graduated from Kennesaw State University in Kennesaw, Georgia. Currently she is writing and editing mathematics books for American Book Company where she has coauthored several books including *Mastering the Grade 8 Common Core in Mathematics* and *Passing the GPS Geometry End-of-Course-Test.* Mary Anne has always enjoyed and excelled in mathematics and plans to attend graduate school in the future.

ABOUT THE REVIEWER

Angela Redman has a B.S. degree in Mathematics Education from Oklahoma State University (1990). She teaches math and is the department chair at Norman North High School in Norman, OK. She has taught math since 1990; ten years spent at the middle school level and twelve years now at the high school level. She's taught seventh grade and eighth grade math, pre-algebra, Algebra 1, Plane Geometry, Intermediate Algebra, and Algebra 2. She has served on various district committees involved with textbook selection, benchmark testing, and curriculum revision.

This book is interactive!

Augmented Reality is an exciting new technology that allows you to interact with printed material from Android and iOS smartphones and tablets. We have implemented this technology in this book to enhance your students' learning experience in a more visual way!

 Use your smartphone or tablet to scan the QR Code to the left and download the AR app for this book.

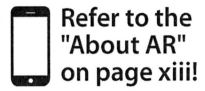 Refer to the "About AR" on page xiii!

When you see the icon on the right throughout the book, refer back to this page to experience your AR content.

American Book Company

The Standards Experts

AUGMENTED REALITY

To learn how to use ABC's Augmented Reality, scan the QR Code to the left with your smartphone or tablet to watch a tutorial video.

You can also visit our web site at **americanbookcompany.com/AugmentedReality** to view the tutorial video.

Below is a list of the locations of the AR content exercises in this book.

Examples

Chapter 1
Page 4 – Example 2
Page 7 – Example 9

Chapter 2
Page 16 – Example 4

Chapter 3
Page 28 – Example 5

Chapter 4
Page 37 - Example 1
Page 38 – Example 2

Chapter 5
Page 57 – Example 14

Chapter 6
Page 74 – Example 8

Chapter 7
Page 82 – Example 5

Chapter 8
Page 98 – Examples

Chapter 9
Page 105 – Example 1

Chapter 10
Page 128 – Example 1

Chapter 11
Page 145 – Example 10

Chapter 12
Page 158 – Example 1

Chapter 13
Page 179 – Example

Chapter 14
Page 190 – Example 5

Chapter 15
Page 199 – Example 2

Activities

1) River/Rabbit Game (Circle Geometry)
2) Car Game (Solving Equations, Simplifying)

Chapter 1
Integers

This chapter covers the following CC Grade 7 standards:

	Content Standards
Expressions and Equations	7.NS.1, 7.NS.2

1.1   Integers (DOK 1)

Whole numbers $= \{0, 1, 2, 3, 4, 5, ...\}$

For most things in life, whole numbers are all we need to use. However, when a checking account falls below zero or the temperature falls below zero, we need a way to express that. We use a negative sign, which looks exactly like a subtraction sign. It is used in front of a number to show that the number is below zero. All the negative whole numbers and positive whole numbers plus zero make up the set of integers.

Integers $= \{..., -4, -3, -2, -1, 0, 1, 2, 3, 4, ...\}$

1.2   Absolute Value (DOK 1)

The absolute value of a number is the distance the number is from zero on the number line.

The absolute value if 6 is written $|6|$. $|6| = 6$
The absolute value of -6 is written $|-6|$. $|-6| = 6$

Both 6 and -6 are the same distance, 6 spaces, from zero so their absolute value is the same: 6.

Simplify the following absolute value problems. (DOK 1)

1. $|9| = $ _____

2. $-|5| = $ _____

3. $|-25| = $ _____

4. $-|-12| = $ _____

5. $-|64| = $ _____

6. $|-2| = $ _____

7. $-|-3| = $ _____

8. $|-4| - |3| = $ _____

9. $|-8| - |-4| = $ _____

10. $|5| + |-4| = $ _____

11. $|-2| + |6| = $ _____

12. $|10| + |8| = $ _____

13. $|-2| + |4| = $ _____

14. $|-3| + |-4| = $ _____

15. $|7| - |-5| = $ _____

1.3   Opposite Numbers (DOK 1)

The **opposite** of a number is the negative of that number. For example, the opposite of 8 is -8. To find the opposite of a negative number, you also take the negative of that number. For example, to find the opposite of -2, take the negative of -2, $-(-2) = 2$. The opposite of -2 is 2. The opposite of a number is the same number, but different sign. The only number that does not have an opposite is 0.

If the number is positive, its opposite will be negative.
If the number is negative, its opposite will be positive.

Find the opposite of the numbers below. (DOK 1)

1. 3

2. $\dfrac{1}{5}$

3. -19

4. 45

5. 1.01

6. $-\dfrac{2}{3}$

7. -2.4

8. $-3\frac{1}{2}$

9. 14

10. $\dfrac{5}{9}$

11. -62

12. 7

13. -1000

14. $-\dfrac{7}{4}$

15. 23

16. 1

17. 7

18. $\dfrac{6}{7}$

19. -2.3

20. $-5\frac{1}{3}$

1.4 Word Problems Using Opposites (DOK 2)

When you combine (add) a number and its opposite, the result is zero. Opposite numbers are used in situations describing the gain or loss of money, rising and falling temperatures, gaining or losing yards in a football game, the charge on an atom, and more. Key words like more, increased, up, forward, and earned indicate that a positive number should be used. Key words like less, negative, decreased, down, back, and spent indicate that a negative number should be used in a problem. Often in these problems you will have to solve for the net change in something. Net change means the final, or overall, change.

Example 1: Sami earned $10 for mowing his neighbor's lawn. The next day, he spent $10 on a book he wants to read. What was Sami's net gain of money?

Step 1: Sami earned $10, so he has $+10$ dollars.

Step 2: Sami then spent $10, so he has -10 dollars.

Step 3: In total, Sami has $10 + (-10) = 0$ dollars.

Answer: Sami's net gain is 0.

Solve the following problems. (DOK 2)

1. In chemistry, an atom's protons contain positive charge, and an atom's electrons contain negative charge. An atom's overall charge is the sum of its positive and negative charges. An atom of silicon has 14 protons and 14 electrons. What is the overall charge on the atom?

2. A hiker climbed up a mountain that is $1,432$ ft tall. He then climbed down the same mountain. What is the hiker's net elevation gain?

3. Jessica earned $20 then spent $13 on a set of paints and $7 on paper and brushes. What is Jessica's net money gain?

4. In one day, the temperature rose $16°$ by midday and then fell $16°$ by nightfall. What was the net change in temperature?

1.5   Adding Integers (DOK 1)

First, we will see how to add integers on the number line; then, we will learn rules for working the problems without using a number line.

Example 2:   Add: $(-3) + 7$

Refer to the "About AR" on page xi!

Step 1:   The first integer in the problem tells us where to start.
Find the first integer, -3, on the number line.

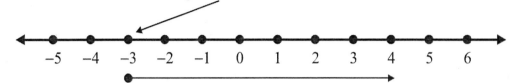

Step 2: $(-3) + 7$ The second integer in the problem, $+7$, tells us the direction to go, positive (toward positive numbers), and how far, 7 places.
$(-3) + 7 = 4$

Example 3:   Add: $(-2) + (-3)$

Step 1:   Find the first integer, (-2), on the number line.

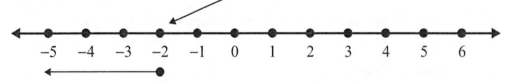

Step 2: $(-2) + (-3)$ The second integer in the problem, (-3), tells us the direction to go, negative (toward the negative numbers), and how far, 3 places.
$(-2) + (-3) = (-5)$

Solve the problems below using this number line. (DOK 1)

1. $2 + (-3)$

2. $4 + (-2)$

3. $(-3) + 7$

4. $(-4) + 4$

5. $(-1) + 5$

6. $(-1) + (-4)$

7. $3 + 2$

8. $(-5) + 8$

9. $3 + (-7)$

10. $(-2) + (-2)$

11. $6 + (-7)$

12. $2 + (-5)$

13. $(-5) + 3$

14. $(-6) + 7$

15. $(-3) + (-3)$

16. $(-8) + 6$

17. $(-2) + 6$

18. $(-4) + 8$

19. $(-7) + 4$

20. $(-5) + 8$

21. $-2 + (-2)$

22. $8 + (-6)$

23. $5 + (-3)$

24. $1 + (-8)$

1.6   Rules for Adding Integers with the Same Signs (DOK 1)

To add integers without using the number line, use these simple rules:

> 1.   **Add the numbers together.**
> 2.   **Give the answer the same sign.**

Example 4: $(-2) + (-5) =$

Both integers are negative. To find the answer, add the numbers together $(2 + 5)$, and give the answer a negative sign.

$(-2) + (-5) = (-7)$

Example 5: $3 + 4 =$

Both integers are positive, so the answer is positive.

$3 + 4 = 7$

NOTE: Sometimes positive signs are placed in front of positive numbers. For example $3 + 4 = 7$ may be written $(+3) + (+4) = +7$. Positive signs are optional. If a number has no sign, it is considered positive.

Solve the problems below using the rules for adding integers with the same signs. (DOK 1)

1. $(-18) + (-4)$
2. $(-12) + (-3)$
3. $(-2) + (-7)$
4. $(22) + (11)$
5. $(-7) + (-6)$
6. $13 + 12$
7. $16 + 11$
8. $(-9) + (-8)$
9. $8 + 4$
10. $(-4) + (-7)$

11. $(-15) + (-5)$
12. $(+7) + (+4)$
13. $(-4) + (-2)$
14. $(-15) + (-1)$
15. $(-8) + (-12)$
16. $6 + 9$
17. $9 + 7$
18. $(-9) + (-7)$
19. $(-14) + (-6)$
20. $(6) + (+19)$

21. $(-11) + (-7)$
22. $(+8) + (+6)$
23. $(+5) + 7$
24. $(-4) + (-9)$
25. $(2) + (8)$
26. $(+18) + 5$
27. $14 + (+7)$
28. $(-11) + (-19)$
29. $13 + (+11)$
30. $(-8) + (-21)$

1.7   Rules for Adding Integers with Opposite Signs (DOK 1)

1. **Ignore the signs and find the difference. (Find the absolute value of each integer.)**
2. **Give the answer the sign of the larger number.**

Example 6: $(-4) + 6 =$

   Step 1: $|(-4)| = 4$ and $|6| = 6$

   Step 2:   To find the difference, take the larger number minus the smaller: $6 - 4 = 2$. Looking back at the original problem, the larger number, 6, is positive, so the answer is positive. $(-4) + 6 = 2$

Example 7: $3 + (-7) =$

   Step 1: $|3| = 3$ and $|(-7)| = 7$

   Step 2:   Find the difference. $7 - 3 = 4$. Looking at the problem, the larger number, 7, is a negative number, so the answer is negative. $3 + (-7) = (-4)$

Solve the problems below using the rules for adding integers with opposite signs. (DOK 1)

1.	$(-4) + 8$	11.	$(-11) + 1$	21.	$-14 + 8$
2.	$-10 + 12$	12.	$(-12) + 8$	22.	$-11 + 15$
3.	$9 + (-3)$	13.	$-14 + 9$	23.	$(-8) + 16$
4.	$(+3) + (-3)$	14.	$14 + (-11)$	24.	$2 + (-15)$
5.	$+8 + (-7)$	15.	$(-20) + 12$	25.	$-2 + 8$
6.	$(-5) + (+12)$	16.	$-19 + 21$	26.	$(-5) + 15$
7.	$-14 + (+7)$	17.	$-4 + 18$	27.	$2 + (-11)$
8.	$15 + (-3)$	18.	$3 + (-6)$	28.	$-3 + 7$
9.	$7 + (-8)$	19.	$4 + (-10)$	29.	$4 + (-12)$
10.	$6 + (-12)$	20.	$(-2) + 8$	30.	$-12 + 5$

Solve the mixed addition problems below using the rules for adding integers. (DOK 1)

31.	$-7 + 8$	37.	$8 + (-5)$	43.	$(-7) + (+10)$
32.	$5 + 6$	38.	$(-6) + 13$	44.	$(+4) + 11$
33.	$(-2) + (-6)$	39.	$(-9) + (-12)$	45.	$11 + 6$
34.	$3 + (-5)$	40.	$(-7) + (+12)$	46.	$-4 + (-10)$
35.	$(-7) + (-9)$	41.	$+8 + (-9)$	47.	$(+6) + (+2)$
36.	$14 + 9$	42.	$(-13) + (-18)$	48.	$1 + (-17)$

1.8   Subtracting Integers (DOK 1)

The easiest way to subtract integers is to change the problem to an addition problem and follow the rules you already know. Use the following rules for subtracting integers:

1.	**Change the subtraction sign to addition.**
2.	**Change the sign of the second number to the opposite sign.**

Example 8: $-6 - (-2) =$

Change the subtraction sign to addition and -2 to 2.

$$-6 - (-2) = -6 + 2 = (-4)$$

Example 9: $5 - 6 =$

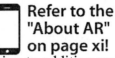

Refer to the "About AR" on page xi!

Change the subtraction sign to addition and 6 to -6.

$$5 - 6 = 5 + (-6) = (-1)$$

Solve the following problems using the appropriate rules. (DOK 1)

1.	$(-3) - 8$	7.	$(-5) - (-13)$	13.	$(-7) - 4$
2.	$5 - (-9)$	8.	$6 - (-7)$	14.	$1 - (-9)$
3.	$8 - (-5)$	9.	$8 - (-6)$	15.	$(-5) - 12$
4.	$(-2) - (-6)$	10.	$(-2) - (-2)$	16.	$(-1) - 9$
5.	$8 - (-9)$	11.	$(-3) - 7$	17.	$6 - (-7)$
6.	$(-4) - (-1)$	12.	$(-4) - 8$	18.	$(-8) - (-12)$

Solve the addition and subtraction problems. (DOK 1)

19.	$4 - (-2)$	27.	$(-5) + 3$	35.	$30 + (-15)$
20.	$(-3) + 7$	28.	$6 - (-7)$	36.	$-40 - (-5)$
21.	$(-4) + 14$	29.	$(-4) + 8$	37.	$25 - 50$
22.	$(-1) - 5$	30.	$(-4) - 11$	38.	$-13 + 12$
23.	$(-1) + (-4)$	31.	$(-5) + 8$	39.	$(-21) - (-1)$
24.	$(-12) + (-2)$	32.	$2 - (-2)$	40.	$62 - (-3)$
25.	$0 - (-6)$	33.	$(-8) + 9$	41.	$(-16) + (-2)$
26.	$2 - (-5)$	34.	$0 + (-10)$	42.	$(-25) + 5$

1.9   Multiplying Integers (DOK 1)

You are probably used to seeing multiplication written with a "×" sign, but multiplication can be written two other ways. A "·" between numbers means the same as "×", and parentheses () around a number without a "×" or a "·" also means to multiply.

Examples: $2 \times 3 = 6$ or $2 \cdot 3 = 6$ or $(2)(3) = 6$

All of these mean the same thing, multiply.

1.10   Dividing Integers (DOK 1)

Division is commonly indicated in two ways: with a "÷" or in the form of a fraction.

Example: $6 \div 3 = 2$ means the same thing as $\dfrac{6}{3} = 2$

1.11   Rules for Multiplying and Dividing Integers (DOK 1)

1.   If the numbers have the same sign, the answer is positive.
2.   If the numbers have different signs, the answer is negative.

Examples: $6 \cdot 8 = 48$ $(-6) \cdot 8 = (-48)$ $(-6) \cdot (-8) = 48$

$48 \div 6 = 8$ $(-48) \div 6 = (-8)$ $(-48) \div (-6) = 8$

Solve the problems below using the rules for multiplying and dividing integers. (DOK 1)

1. $(-4) \div 2$

2. $12 \div (-3)$

3. $\dfrac{(-14)}{(-2)}$

4. $-15 \div 3$

5. $(-3) \cdot (-7)$

6. $(-1) \cdot (5)$

7. $-1 \cdot (-4)$

8. $2(-5)$

9. $3 \cdot (-7)$

10. $(-12) \cdot (-2)$

11. $\dfrac{(-18)}{(-6)}$

12. $21 \div (-7)$

13. $-5 \cdot 3$

14. $(-6)(7)$

15. $(-5) \cdot 8$

16. $\dfrac{-12}{6}$

17. $8(-4)$

18. $1 \cdot (-8)$

19. $(-7) \cdot (-4)$

20. $(-2) \div (-2)$

21. $\dfrac{18}{(-6)}$

1.12   Going Deeper into Integers (DOK 3)

The example below is a review of the rules of the order of operations. Follow the phrase **Please Excuse My Dear Aunt Sally** as a memory aid.

Example 10: $24 \div 3\,(-5 - 3) + 2^2 - 5$

Please	First, do the operation inside the **parentheses**. $-5 - 3 = -8$	$24 \div 3 \times -8 + 2^2 - 5$
Excuse	Second, evaluate **exponents**. $2^2 = 4$ so now we have	$24 \div 3 \times -8 + 4 - 5$
My Dear	Third, **multiply** and **divide** next in order from left to right.	$24 \div 3 = 8;\ 8 \times -8 = -64$
Aunt Sally	Last, **add** and **subtract** in order from left to right.	$-64 + 4 - 5 = -65$

Use the information in this chapter about integers and the rules of the order of operations to solve the multi-step problems below. Show your work for each step. (DOK 3)

1. $(-2)^2 + 3(8 - 12)$

2. $100 - (-10) + (5 + 5)^2$

3. $(-81 \div 9) + (-12 \times 2) - 2$

4. $7 \times (3)^2 + (-2 - 2)$

5. $(36 \div 12)^2 + (-8 \div 2)$

6. $15 - 5(2 \times 3) - (-15)$

7. $(-4 \times 4) + (-80 \div 20) + 4$

8. $(-3)^2 - (5 \times 7) + 20$

9. $(1,000 \div 10) - (-10 \times 10)$

10. $37 + (-4)^2 + (-90 \div 3)$

11. $15 \times 4 - (-4 \times 3) + 2$

12. $(-7)^2 + (-55 \div 11) + (-8)$

13. $(3 + 2)^2 - (-5 \times 2)$

14. $2 \times (27 \div 3) + (-3 \times 3)$

15. $-14 \times 2 + (-7 + 2)$

16. $(2 \times 2)^2 - (-2 \times -2) + 2$

17. Explain how you would simplify the equation $(-1)^2 + b(c - d)$, where a, b, c, and d are all integers greater than zero. Which step would come first? Which step would be last? How do you know? Justify your explanation.

Chapter 1 Review

Simplify. (DOK 1)

1. $(-6) + 13$

2. $(-3) + (-9)$

3. $(-4) \cdot 4$

4. $(-18) \div 3$

5. $(-1) - 5$

6. $(-1) \cdot (-4)$

7. $3 + (-5)$

8. $6 + (-5)$

9. $\dfrac{12}{(-3)}$

10. $2 + (-5)$

11. $\dfrac{(-24)}{(-6)}$

12. $(-5) + 3$

13. $(-6) - 7$

14. $(-33) \div (-11)$

15. $(-21)(-3)$

16. $(-7) + (-14)$

17. $(-5) - 8$

18. $1(-8)$

19. $(-2) \cdot (-2)$

20. $8 + (-6)$

21. $\dfrac{-14}{7}$

22. $(7) \cdot (-2)$

23. $(10)(4)$

24. $24 \div (-4)$

25. $6(-5)$

26. $(-9) - (-12)$

27. $36 \div 12$

Solve the following absolute value problems. (DOK 1)

28. $|4|$

29. $|-6|$

30. $|-3| + |7|$

31. $|8| - |-5|$

Find the opposite of the numbers below. (DOK 1)

32. -42

33. $\dfrac{1}{16}$

34. $-\dfrac{5}{8}$

35. 26.1

Solve the following problems. (DOK 2)

36. A football team gained 4 yards in one play. In the next play, the team lost 4 yards. What is the team's net yard gain?

37. Between 8 a.m. and 3 p.m. on a summer day, the outdoor temperature increased by $23°$. Between 3 p.m. and 10 p.m., the temperature decreased by $23°$. What was the net change in temperature from 8 a.m. to 10 p.m.?

Solve the multi-step problems below using the rules of order of operations. Show your work for each step. (DOK 3)

38. $(15 \div 5)^2 + (-6 \times 2)$

39. $18 \div (-2 \times -3) - 3$

Chapter 1 Test

1 $7 + (-5) =$

A -2
B 2
C -12
D 12

2 $(-3) + 23 =$

A -26
B -20
C 26
D 20

3 $(-13) + 11 =$

A -2
B -24
C -143
D 2

4 $1337 - (-1337) =$

A 2674
B 0
C -2674
D $-1,787,569$

5 $(-81) - 33 =$

A 48
B -48
C -114
D 114

6 $(-17) - (-37) =$

A -54
B 20
C -20
D 54

7 $24 \div (-6) =$

A -18
B -4
C 30
D 4

8 $(-169) \div (-13) =$

A -182
B -13
C 13
D -156

9 $(-10) \cdot (-7) =$

A 70
B -13
C 13
D -156

10 $(3) \times (-12) =$

A -9
B 15
C -4
D -36

11 What is the opposite of -9?

    A 9

    B $\dfrac{1}{9}$

    C -9

    D $-\dfrac{1}{9}$

                         (DOK 1)

12 Johnny receives a weekly allowance of $7. He has been saving his allowance for three weeks in order to buy a skateboard for $21. After he buys the skateboard, what will be Johnny's net gain in money?

    A $7

    B $0

    C $14

    D $21

                         (DOK 2)

13 A molecule of water contains two hydrogen atoms and one oxygen atom. Each hydrogen atom contains one proton and one electron. An oxygen atom contains 8 protons and 8 electrons. Each proton contains one unit of positive charge, and each electron contains one unit of negative charge. What is the overall charge on a water molecule?

    A $+2$

    B -2

    C 8

    D 0

                         (DOK 2)

14 Which of the sentences below is <u>not</u> true?

    A $2 - (-2) = 0$

    B $-2 - 2 = -4$

    C $-2 - (-2) = 0$

    D $2 - 2 = 4$

                         (DOK 3)

15 Which of the sentences below is true?

    A $-6 \times (-3) = -18$

    B $-6 \times 3 = -18$

    C $-6 \div 3 = 2$

    D $-6 \div (-3) = -2$

                         (DOK 3)

16 Solve: $(4 + 2)^2 - (-30)$

    A 36

    B 66

    C -36

    D -66

                         (DOK 3)

17 Solve: $-45 - (9 \div 3)^2 + (-3)$

    A -39

    B 57

    C -57

    D 39

                         (DOK 3)

Chapter 2
Rational Numbers

This chapter covers the following CC Grade 7 standards:

	Content Standards
Expressions and Equations	7.NS.1, 7.NS.2

2.1   Rational Numbers (DOK 1)

Rational numbers are numbers that can be written as a fraction. The division of two integers always equals a rational number. Rational numbers include:

1 whole numbers

2 zero

3 positive and negative integers

4 positive and negative fractions

5 positive and negative terminating and repeating decimals

(Reminder: Whole numbers are fractions when written over 1. Example: 4 can be written as $\frac{4}{1}$. Also, calculators will often truncate large numbers.)

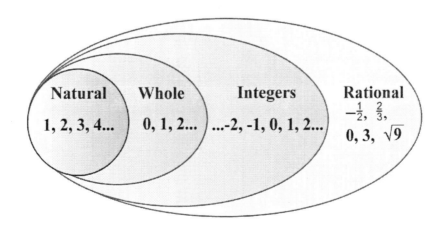

2.2   Comparing Fractions (DOK 2)

Compare fractions using the greater than (>), less than (<), and equal to (=) signs.

Example 1:   Compare $\frac{3}{4}$ and $\frac{5}{8}$

Step 1:   Find the lowest common denominator (the smallest number that both denominators will divide into evenly.) The lowest common denominator is 8.

Step 2:   Change fourths to eighths by multiplying three fourths by two halves,

$$\frac{2 \times 3}{2 \times 4} = \frac{6}{8}.$$

Step 3: $\frac{6}{8} > \frac{5}{8}$ Since the denominators are the same, the fraction with the larger number is greater, therefore, $\frac{3}{4} > \frac{5}{8}$.

Example 2:   Compare the mixed numbers $1\frac{3}{5}$ and $1\frac{2}{3}$.

Step 1:   Change the mixed numbers to improper fractions.

$$1\frac{3}{5} = \frac{8}{5} \text{ and } 1\frac{2}{3} = \frac{5}{3}$$

Step 2:   Find the lowest common denominator for the improper fractions. The lowest common denominator is 15.

Step 3:   Change fifths to fifteenths and thirds to fifteenths, $\frac{3 \times 8}{3 \times 5} = \frac{24}{15}$ and $\frac{5 \times 5}{5 \times 3} = \frac{25}{15}$.

Step 4: $\frac{24}{15} < \frac{25}{15}$, therefore, $1\frac{3}{5} < 1\frac{2}{3}$.

Fill in the box with the correct sign (>, <, or =). (DOK 2)

1. $\frac{7}{9} \square \frac{7}{8}$

2. $\frac{6}{7} \square \frac{5}{6}$

3. $\frac{4}{6} \square \frac{5}{7}$

4. $\frac{3}{10} \square \frac{4}{13}$

5. $\frac{5}{8} \square \frac{4}{11}$

6. $\frac{5}{8} \square \frac{4}{7}$

7. $\frac{9}{10} \square \frac{8}{13}$

8. $\frac{2}{13} \square \frac{1}{10}$

9. $\frac{4}{9} \square \frac{3}{5}$

10. $\frac{2}{6} \square \frac{4}{5}$

11. $\frac{7}{12} \square \frac{6}{11}$

12. $\frac{3}{11} \square \frac{5}{12}$

2.3   Comparing Decimals (DOK 2)

A **decimal** is the point after a whole number. A decimal is commonly used in monetary notation, such as $12.54, where the dollars are to the left of the decimal and the cents are to the right of the decimal. A decimal is also used to express a whole number and a fraction, such as one and a quarter pies expressed as 1.25 pies. Decimals preceded by a zero represent numbers less than one, such as a quarter of a pie expressed as 0.25 pies. Be sure to include a leading zero to the left of the decimal for numbers less than one. A decimal is read as "point"; 1.25 is read as "one point two five" (or "one point twenty-five"). The number 1.25 can also be read as "one <u>and</u> twenty-five hundredths."

The number below is **1,863.681726**. Study the place value names.

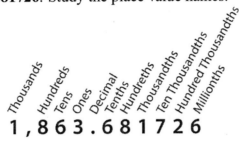

$$1,863.681726$$

Example 3:	Compare 4.1 and 4.11.
Step 1:	Arrange numbers with decimal points directly under each other.
	4.1
	4.11
Step 2:	Fill in with zeros so they all have the same number of digits after the decimal point. **Read the numbers as if the decimal points were not there.**
	4.10
	4.11
Answer:	4.1 < 4.11

Fill in the box with the correct sign (>, <, or =). (DOK 2)

1.   7.2 ☐ 7.02

2.   0.30 ☐ 0.3

3.   9.44 ☐ 9.4

4.   21.01 ☐ 21.11

5.   6.22 ☐ 6.20

6.   17.1001 ☐ 17.1011

7.   8.55 ☐ 8.555

8.   29.06 ☐ 29.060

9.   40.99 ☐ 40.9

10.   97.08 ☐ 97.18

11.   0.21 ☐ 0.211

12.   36.66 ☐ 36.606

2.4   Changing Fractions to Decimals (DOK 1)

Example 4:   Change $\frac{1}{8}$ to a decimal.

Refer to the "About AR" on page xi!

Step 1:   To change a fraction to a decimal, divide the top number by the bottom number.

$8\overline{)1}$

Step 2:   Add a decimal point and a 0 after the 1 and divide.

$$\begin{array}{r} 0.1 \\ 8\overline{)1.0} \\ -8 \\ \hline 2 \end{array}$$

Step 3:   Continue adding 0's and dividing until there is no remainder.

$$\begin{array}{r} 0.125 \\ 8\overline{)1.000} \\ -8 \\ \hline 20 \\ -16 \\ \hline 40 \\ -40 \\ \hline 0 \end{array}$$

In some problems, the number after the decimal point begins to repeat. Take, for example, the fraction $\frac{4}{11}$. $4 \div 11 = 0.363636$, and the 36 keeps repeating forever. To show that the 36 repeats, write a bar above the numbers that repeat, $0.\overline{36}$.

Change the following fractions to decimals. (DOK 1)

1. $\frac{4}{5}$ 6. $\frac{5}{8}$ 11. $\frac{5}{11}$ 16. $\frac{3}{8}$

2. $\frac{2}{3}$ 7. $\frac{5}{6}$ 12. $\frac{1}{9}$ 17. $\frac{3}{16}$

3. $\frac{1}{2}$ 8. $\frac{1}{6}$ 13. $\frac{7}{9}$ 18. $\frac{3}{4}$

4. $\frac{5}{9}$ 9. $\frac{3}{5}$ 14. $\frac{9}{10}$ 19. $\frac{8}{9}$

5. $\frac{1}{10}$ 10. $\frac{7}{10}$ 15. $\frac{1}{4}$ 20. $\frac{5}{12}$

If there is a whole number with a fraction, write the whole number to the left of the decimal point. Then, change the fraction to a decimal.

Example 5: $4\frac{1}{10} = 4.1$ $\qquad$ $16\frac{2}{3} = 16.\overline{6}$ $\qquad$ $12\frac{7}{8} = 12.875$

Change the following mixed numbers to decimals. (DOK 1)

1. $5\frac{2}{3}$ 5. $30\frac{1}{3}$ 9. $6\frac{4}{5}$ 13. $7\frac{1}{4}$ 17. $10\frac{1}{10}$

2. $8\frac{5}{11}$ 6. $3\frac{1}{2}$ 10. $13\frac{1}{2}$ 14. $12\frac{1}{3}$ 18. $20\frac{2}{5}$

3. $15\frac{3}{5}$ 7. $1\frac{7}{8}$ 11. $12\frac{4}{5}$ 15. $1\frac{5}{8}$ 19. $4\frac{9}{10}$

4. $13\frac{2}{3}$ 8. $4\frac{9}{100}$ 12. $11\frac{5}{8}$ 16. $2\frac{3}{4}$ 20. $5\frac{4}{11}$

2.5 Changing Decimals to Fractions (DOK 1)

Example 6: Change 0.25 to a fraction.

Step 1: Copy the decimal without the point. This will be the top number of the fraction.
$$\frac{25}{}$$

Step 2: The bottom number is a 1 with as many 0's after it as there are digits in the top number. $\qquad$ $\dfrac{25 \;\leftarrow\; \text{Two digits}}{100 \;\leftarrow\; \text{Two 0's}}$

Step 3: You then need to simplify the fraction. $\qquad$ $\dfrac{25}{100} = \dfrac{1}{4}$

Examples: $0.2 = \dfrac{2}{10} = \dfrac{1}{5}$ $\qquad$ $0.65 = \dfrac{65}{100} = \dfrac{13}{20}$ $\qquad$ $0.125 = \dfrac{125}{1000} = \dfrac{1}{8}$

Change the following decimals to fractions. (DOK 1)

1. 0.55   5. 0.75   9. 0.71   13. 0.35

2. 0.6   6. 0.82   10. 0.32   14. 0.96

3. 0.12   7. 0.3   11. 0.56   15. 0.125

4. 0.9   8. 0.42   12. 0.24   16. 0.375

Example 7:   Change 14.28 to a mixed number.

Step 1:   Copy the portion of the number that is whole.

14

Step 2:   Change 0.28 to a fraction.

$$14\frac{28}{100}$$

Step 3:   Simplify the fraction.

$$14\frac{28}{100} = 14\frac{7}{25}$$

Change the following decimals to mixed numbers. (DOK 1)

1.   7.125

2.   99.5

3.   2.13

4.   5.1

5.   16.95

6.   3.625

7.   4.42

8.   15.84

9.   6.7

10.   45.425

11.   15.8

12.   8.16

13.   13.9

14.   32.65

15.   17.25

16.   9.82

2.6   Graphing Rational Numbers on a Number Line (DOK 2)

Improper fractions, decimal values, and all other rational numbers can be plotted on a number line. Study the examples below.

Example 8:   Where would $\frac{4}{3}$ fall on the number line below?

Step 1:   Convert the improper fraction to a mixed number. $\frac{4}{3} = 1\frac{1}{3}$

Step 2: $1\frac{1}{3}$ is $\frac{1}{3}$ of the distance between the numbers 1 and 2. Estimate this distance by dividing the distance between points 1 and 2 into thirds. Plot the point at the first division.

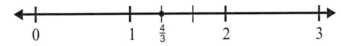

Example 9:   Plot the value of -1.75 on the number line below.

Step 1:   Convert the value -1.75 to a mixed number. $-1.75 = -1\frac{3}{4}$

Step 2: $-1\frac{3}{4}$ is $\frac{3}{4}$ of the distance between the numbers -1 and -2. Estimate this distance by dividing the distance between points -1 and -2 into fourths. Plot the point at the third division.

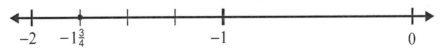

Example 10:   Plot the value of $3.5 \div 2$ on the number line below.

Step 1:   Figure the value of $3.5 \div 2$. $3.5 \div 2 = 1.75$ or $1\frac{3}{4}$.

Step 2:   Plot 1.75.

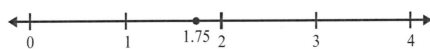

Plot and label the following values on the number lines given below. (DOK 2)

1. $A = \dfrac{5}{4}$ $\qquad\qquad$ $B = \dfrac{12}{5}$ $\qquad\qquad$ $C = \dfrac{2}{3}$ $\qquad\qquad$ $D = -\dfrac{3}{2}$

2. $E = 1.4$ $\qquad\qquad$ $F = -2.25$ $\qquad\qquad$ $G = -0.6$ $\qquad\qquad$ $H = 0.625$

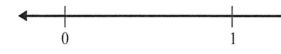

3. $I = 0.25$ $\qquad\qquad$ $J = 0.9$ $\qquad\qquad$ $K = 1.9$ $\qquad\qquad$ $L = 2.6$

Match the correct value for each point on the number line below. (DOK 2)

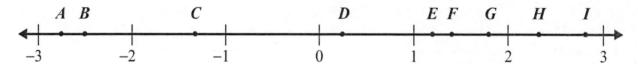

4. $1.8 = $ _____

5. $\dfrac{7}{3} = $ _____

6. $1\dfrac{1}{3} = $ _____

7. $-\dfrac{5}{2} = $ _____

8. $-2.75 = $ _____

9. $-\dfrac{4}{3} = $ _____

10. $2\dfrac{4}{5} = $ _____

11. $\dfrac{6}{5} = $ _____

12. $0.25 = $ _____

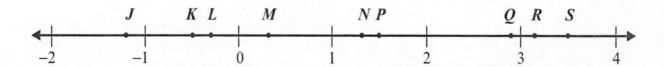

13. $\dfrac{19}{6} = $ _____

14. $-0.5 = $ _____

15. $\dfrac{5}{4} = $ _____

16. $\dfrac{1}{3} = $ _____

17. $1.5 = $ _____

18. $-0.3 = $ _____

19. $-\dfrac{6}{5} = $ _____

20. $3\dfrac{1}{2} = $ _____

21. $2.9 = $ _____

2.7 Comparing Rational Numbers (DOK 2)

When comparing numbers, use the greater than (>), less than (<), and the equal to (=) signs. The simplest way to compare numbers that are in different notations, like percent, decimals, and fractions, is to change all of them to one notation. <u>Decimals are the easiest to compare.</u>

Example 11: Which is larger: $1\frac{1}{4}$ or 1.3?

Answer: Change $1\frac{1}{4}$ to a decimal. $\frac{1}{4} = 0.25$, so $1\frac{1}{4} = 1.25$, which is smaller than 1.3.

Example 12: Which is smaller: 0.6 or $\frac{2}{3}$?

Answer: Change $\frac{2}{3}$ to a decimal.

$$\frac{2}{3} = 0.\overline{66}$$

0.6 is smaller than $0.\overline{66}$, so $0.6 < \frac{2}{3}$

Fill in each box with the correct sign. (DOK 2)

1. $23.4 \ \square \ 23\frac{1}{2}$

2. $\frac{17}{100} \ \square \ -0.17$

3. $\frac{3}{8} \ \square \ 0.38$

4. $0.25 \ \square \ \frac{2}{10}$

5. $23.4 \ \square \ 23\frac{1}{4}$

6. $-\frac{1}{7} \ \square \ 0.14$

7. $13.95 \ \square \ 13\frac{8}{9}$

8. $-\frac{4}{5} \ \square \ -0.4$

9. $1.25 \ \square \ \frac{3}{2}$

10. $\frac{12}{4} \ \square \ -3$

11. $0.06 \ \square \ \frac{1}{16}$

12. $-1.\overline{33} \ \square \ -\frac{4}{3}$

13. Will this inequality below be <u>true</u> when $x = \frac{2}{5}$?

$$0.23 \le x < 0.4$$

14. Will this inequality below be <u>true</u> when $n = 0.15$?

$$\frac{1}{7} \le n \le \frac{1}{6}$$

2.8   Going Deeper into Rational Numbers (DOK 3)

Some of the comparisons below are incorrect. If they are incorrect, change the comparison sign to make the problem correct. If they are correct, simply write C = correct. (DOK 3)

1. $\frac{2}{3} > 0.875$

2. $1.45 < 1\frac{4}{5}$

3. $\frac{16}{3} = 5.\overline{3}$

4. $2.375 = \frac{17}{8}$

5. $\frac{1}{4} > 0.24$

6. $0.91 < \frac{9}{10}$

7. $-10.1 = -\frac{10}{101}$

8. $\frac{4}{5} > 0.8$

9. $1.1 > 1\frac{1}{4}$

10. $\frac{11}{10} < 1.01$

11. $-1\frac{1}{4} < -1.15$

12. $\frac{7}{8} > 0.625$

13. $0.85 = \frac{4}{5}$

14. $2\frac{3}{5} > 2.6$

15. $0.5 = \frac{7}{15}$

16. $0.75 > \frac{6}{9}$

17. $1.5 = \frac{2}{3}$

18. $\frac{9}{10} < 0.91$

Answer each question about rational numbers. Explain your answers. (DOK 3)

19. Write a fraction with a denominator of 36 that <u>cannot</u> be simplified. Explain why this is possible.

20. The Cool Boutique and Rachel's Rags are two clothing stores offering sales discounts. The Cool Boutique offers a 33% discount and Rachel's Rags offers a $\frac{1}{3}$ off deal. Which store offers the bigger discount? Explain why.

21. Convert $\frac{41}{99}$ into a decimal, then write what you think the decimal equivalent of $\frac{77}{99}$ is. Explain your answer.

22. If $\frac{a}{b}$ is a rational number, what does b have to equal in order for $\frac{a}{b}$ to also be an integer?

Chapter 2 Review

Fill in the box with the correct sign (>, <, or =). (DOK 2)

1. $\dfrac{2}{3} \ \square \ \dfrac{2}{5}$

2. $0.75 \ \square \ \dfrac{7}{8}$

3. $\dfrac{1}{2} \ \square \ \dfrac{3}{6}$

4. $\dfrac{5}{8} \ \square \ 0.75$

5. $-6 \ \square \ -5$

6. $2.1 \ \square \ 2.01$

7. $16.9 \ \square \ 16\frac{9}{10}$

8. $0.42 \ \square \ 0.4$

Change the following numbers from a decimal to a fraction. (DOK 1)

9. 0.55

10. 0.84

11. 0.32

12. 1.35

13. 4.2

Change the following numbers from a fraction to a decimal. (DOK 1)

14. $5\dfrac{3}{25}$

15. $\dfrac{7}{100}$

16. $10\dfrac{2}{3}$

17. $\dfrac{3}{8}$

18. $7\dfrac{3}{4}$

Some of the comparisons below are incorrect. If they are incorrect, change the comparison sign to make the problem correct. If they are correct, simply write C = correct. (DOK 3)

19. $\dfrac{2}{5} > 0.04$

20. $1.05 < 1\dfrac{5}{100}$

21. $\dfrac{21}{2} = 10.1$

22. $1.625 = \dfrac{13}{8}$

23. $\dfrac{1}{5} > 0.19$

24. $0.82 < \dfrac{8}{12}$

Answer each question about rational numbers. Explain your answers. (DOK 3)

25. Write a fraction with a denominator of 12 that <u>cannot</u> be simplified. Explain why this is possible.

26. Jesse wants to buy the cheaper of two packages of cereal that weigh exactly the same amount. One cereal, Oaty-Oats costs \$4.29 and Jesse has a \$1.00 coupon for this box. The other cereal, Wheaty-Wheats is also \$4.29, but the manager has it on special for 25% off. Which box of cereal is the better deal for Jesse? Explain why.

Chapter 2 Test

1 What is the value of $\frac{5}{16}$ in decimal form?

A 0.3125

B 0.3025

C 3.2

D 3.125

(DOK 1)

2 What is the value of 1.875 in fractional form?

A $\frac{7}{5}$

B $\frac{9}{8}$

C $\frac{15}{8}$

D $\frac{7}{6}$

(DOK 1)

3 What sign would you place in the box below to accurately compare the two numbers?

$$0.0625 \ \square \ 0.0615$$

A $>$

B $<$

C $=$

D You cannot compare the two

(DOK 2)

4 What sign would you place in the box below to accurately compare the two fractions?

$$\frac{3}{8} \ \square \ \frac{2}{5}$$

A $<$

B $>$

C $=$

D You cannot compare the two fractions.

(DOK 2)

5 Compare the two numbers below.

$$\frac{5}{7} \ \square \ 0.83$$

A $>$

B $<$

C $=$

D You cannot compare a fraction and a decimal to each other.

(DOK 2)

6 Compare the two numbers below.

$$4.8 \ \square \ 4\frac{4}{5}$$

A $>$

B $<$

C $=$

D You cannot compare a fraction and a decimal to each other.

(DOK 2)

7 Four different kinds of shirts normally cost the same price of $19.99. Shirt A is on sale for 25% off, shirt B is on sale for $5.00 off, shirt C is on sale for "buy one, get one half off", and shirt D is 0.7 of the original price. If Steven wants to buy two shirts of the same kind, which two shirts will give him the best price?

A 2 of shirt A

B 2 of shirt B

C 2 of shirt C

D 2 of shirt D

(DOK 3)

8 Which comparison below is _not_ correct?

A $1.25 > \frac{4}{5}$

B $1.25 = \frac{4}{5}$

C $1.25 < \frac{6}{4}$

D $1.25 = \frac{6}{4}$

(DOK 3)

Chapter 3
Adding and Subtracting Rational Numbers

This chapter covers the following CC Grade 7 standards:

	Content Standards
Expressions and Equations	7.NS.1, 7.NS.3

3.1 Simplifying Improper Fractions (DOK 1)

Example 1: Simplify $\dfrac{21}{4} = 21 \div 4 = 5$ remainder 1

The quotient, 5, becomes the whole number portion of the mixed number.

$$\dfrac{21}{4} = 5\dfrac{1}{4} \quad \text{The remainder, 1, becomes the top number of the fraction.}$$

The bottom number of the fraction always remains the same.

Example 2: Simplify $\dfrac{11}{6}$.

Step 1: $\dfrac{11}{6}$ is the same as $11 \div 6$. $11 \div 6 = 1$ with a remainder of 5.

Step 2: Rewrite as a whole number with a fraction. $1\dfrac{5}{6}$

Simplify the following improper fractions. (DOK 1)

1. $\dfrac{13}{5} =$ _____
2. $\dfrac{11}{3} =$ _____
3. $\dfrac{24}{6} =$ _____
4. $\dfrac{7}{6} =$ _____

5. $\dfrac{19}{6} =$ _____
6. $\dfrac{16}{7} =$ _____
7. $\dfrac{13}{8} =$ _____
8. $\dfrac{9}{5} =$ _____

9. $\dfrac{22}{3} =$ _____
10. $\dfrac{13}{4} =$ _____
11. $\dfrac{15}{2} =$ _____
12. $\dfrac{22}{9} =$ _____

13. $\dfrac{17}{9} =$ _____
14. $\dfrac{27}{8} =$ _____
15. $\dfrac{32}{7} =$ _____
16. $\dfrac{3}{2} =$ _____

17. $\dfrac{7}{4} =$ _____
18. $\dfrac{21}{10} =$ _____
19. $\dfrac{82}{9} =$ _____
20. $\dfrac{44}{6} =$ _____

3.2   Adding Fractions (DOK 1)

Example 3:   Add: $3\frac{1}{2} + 2\frac{2}{3}$

Step 1:   Rewrite the problem vertically, and find a common denominator.
Think: What is the smallest number I can divide 2 and 3 into without a remainder? 6, of course.

$$3\frac{1}{2} = \frac{}{6}$$
$$+2\frac{2}{3} = \frac{}{6}$$

Step 2:   To find the numerator for the top fraction, think: What do I multiply 2 by to get 6? You must multiply the top and bottom numbers of the fraction by 3 to keep the fraction equal. For the bottom fraction, multiply the top and bottom number by 2.

Step 3:   Add whole numbers and fractions, and simplify.

$$3\frac{1}{2} = 3\frac{3}{6}$$
$$+2\frac{2}{3} = 2\frac{4}{6}$$
$$= 5\frac{7}{6} = 6\frac{1}{6}$$

Add and simplify the answers. (DOK 1)

1. $2\frac{3}{4}$ $+5\frac{1}{8}$

2. $2\frac{3}{4}$ $+\frac{7}{8}$

3. $\frac{2}{7}$ $+\frac{4}{5}$

4. $3\frac{1}{7}$ $+1\frac{1}{2}$

5. $10\frac{2}{3}$ $+1\frac{1}{2}$

6. $9\frac{1}{8}$ $+\frac{2}{6}$

7. $7\frac{1}{8}$ $+2\frac{3}{5}$

8. $4\frac{7}{8}$ $+3\frac{1}{9}$

9. $8\frac{5}{6}$ $+3\frac{7}{10}$

10. $1\frac{5}{8}$ $+\frac{3}{4}$

11. $2\frac{5}{12}$ $+1\frac{4}{9}$

12. $\frac{2}{7}$ $+\frac{2}{5}$

13. $2\frac{1}{9}$ $+3\frac{3}{4}$

14. $1\frac{7}{8}$ $+3\frac{3}{12}$

15. $7\frac{1}{4}$ $+3\frac{2}{3}$

16. $4\frac{4}{5}$ $+4\frac{3}{7}$

17. $\frac{7}{12}$ $+4\frac{2}{3}$

18. $5\frac{7}{8}$ $+1\frac{1}{9}$

19. $6\frac{9}{10}$ $+2\frac{7}{10}$

20. $1\frac{1}{9}$ $+7\frac{3}{4}$

21. $2\frac{1}{2}$ $+3\frac{1}{7}$

3.3   Subtracting Mixed Numbers from Whole Numbers (DOK 1)

Example 4:   Subtract: $15 - 3\frac{3}{4}$

Step 1:   Rewrite the problem vertically.

$$\begin{array}{r} 15 \\ -\ 3\frac{3}{4} \\ \hline \end{array}$$

Step 2:   You cannot subtract three-fourths from nothing. You must borrow 1 from 15. You will need to put the 1 in the fraction form. If you use $\frac{4}{4}$ $\left(\frac{4}{4} = 1\right)$, you will be ready to subtract.

$$\begin{array}{r} \overset{4}{\cancel{15}}\frac{4}{4} \\ -\ 3\frac{3}{4} \\ \hline 11\frac{1}{4} \end{array}$$

Subtract. (DOK 1)

1. $\begin{array}{r} 7 \\ -2\frac{3}{5} \\ \hline \end{array}$

2. $\begin{array}{r} 24 \\ -12\frac{1}{2} \\ \hline \end{array}$

3. $\begin{array}{r} 13 \\ -11\frac{2}{3} \\ \hline \end{array}$

4. $\begin{array}{r} 28 \\ -21\frac{5}{8} \\ \hline \end{array}$

5. $\begin{array}{r} 12 \\ -6\frac{1}{8} \\ \hline \end{array}$

6. $\begin{array}{r} 8 \\ -6\frac{3}{4} \\ \hline \end{array}$

7. $\begin{array}{r} 22 \\ -9\frac{1}{2} \\ \hline \end{array}$

8. $\begin{array}{r} 14 \\ -7\frac{1}{5} \\ \hline \end{array}$

9. $\begin{array}{r} 15 \\ -12\frac{2}{9} \\ \hline \end{array}$

10. $\begin{array}{r} 35 \\ -22\frac{7}{9} \\ \hline \end{array}$

11. $\begin{array}{r} 18 \\ -7\frac{5}{6} \\ \hline \end{array}$

12. $\begin{array}{r} 40 \\ -36\frac{11}{13} \\ \hline \end{array}$

13. $\begin{array}{r} 3 \\ -1\frac{1}{3} \\ \hline \end{array}$

14. $\begin{array}{r} 12 \\ -4\frac{3}{8} \\ \hline \end{array}$

15. $\begin{array}{r} 5 \\ -2\frac{1}{8} \\ \hline \end{array}$

16. $\begin{array}{r} 15 \\ -1\frac{15}{16} \\ \hline \end{array}$

17. $\begin{array}{r} 10 \\ -8\frac{4}{5} \\ \hline \end{array}$

18. $\begin{array}{r} 6 \\ -5\frac{1}{7} \\ \hline \end{array}$

19. $\begin{array}{r} 2 \\ -\frac{3}{4} \\ \hline \end{array}$

20. $\begin{array}{r} 37 \\ -24\frac{2}{7} \\ \hline \end{array}$

21. $\begin{array}{r} 2 \\ -\frac{5}{9} \\ \hline \end{array}$

22. $\begin{array}{r} 4 \\ -2\frac{1}{2} \\ \hline \end{array}$

23. $\begin{array}{r} 6 \\ -5\frac{1}{9} \\ \hline \end{array}$

24. $\begin{array}{r} 12 \\ -11\frac{1}{9} \\ \hline \end{array}$

25. $\begin{array}{r} 9 \\ -7\frac{1}{8} \\ \hline \end{array}$

3.4   Subtracting Mixed Numbers with Borrowing (DOK 2)

Example 5:   Subtract: $7\frac{1}{4} - 5\frac{5}{6}$

Refer to the "About AR" on page xi!

Step 1:   Rewrite the problem and find a common denominator.

$$7\frac{1}{4} \quad \begin{array}{c} \times 3 \\ \times 3 \end{array} \quad \rightarrow \quad 7\frac{3}{12}$$

$$-5\frac{5}{6} \quad \begin{array}{c} \times 2 \\ \times 2 \end{array} \quad \rightarrow \quad -5\frac{10}{12}$$

Step 2:   You cannot subtract 10 from 3. You must borrow 1 from the 7. The 1 will be in the fraction form $\frac{12}{12}$ which you must add to the $\frac{3}{12}$ you already have, making $\frac{15}{12}$.

Step 3:   Subtract whole numbers and fractions, and simplify.

$$\begin{array}{r} \overset{6}{7}\overset{15}{\frac{\cancel{3}}{12}} \\ -5\frac{10}{12} \\ \hline 1\frac{5}{12} \end{array}$$

Subtract and simplify. (DOK 2)

1. $\begin{array}{r} 5\frac{1}{8} \\ -3\frac{3}{4} \\ \hline \end{array}$

2. $\begin{array}{r} 8\frac{2}{7} \\ -2\frac{1}{2} \\ \hline \end{array}$

3. $\begin{array}{r} 4\frac{11}{12} \\ -3\frac{1}{3} \\ \hline \end{array}$

4. $\begin{array}{r} 7\frac{4}{6} \\ -4\frac{2}{3} \\ \hline \end{array}$

5. $\begin{array}{r} 15\frac{3}{10} \\ -7\frac{1}{2} \\ \hline \end{array}$

6. $\begin{array}{r} 3\frac{1}{5} \\ -1\frac{2}{15} \\ \hline \end{array}$

7. $\begin{array}{r} 16\frac{5}{12} \\ -12\frac{1}{6} \\ \hline \end{array}$

8. $\begin{array}{r} 20\frac{2}{7} \\ -6\frac{4}{5} \\ \hline \end{array}$

9. $\begin{array}{r} 7\frac{4}{9} \\ -3\frac{1}{3} \\ \hline \end{array}$

10. $\begin{array}{r} 8\frac{4}{5} \\ -3\frac{1}{10} \\ \hline \end{array}$

11. $\begin{array}{r} 18\frac{2}{3} \\ -17\frac{5}{6} \\ \hline \end{array}$

12. $\begin{array}{r} 11\frac{1}{2} \\ -1\frac{7}{8} \\ \hline \end{array}$

13. $\begin{array}{r} 14\frac{1}{2} \\ -2\frac{2}{3} \\ \hline \end{array}$

14. $\begin{array}{r} 11\frac{5}{6} \\ -3\frac{1}{3} \\ \hline \end{array}$

15. $\begin{array}{r} 9\frac{11}{13} \\ -8\frac{2}{13} \\ \hline \end{array}$

16. $\begin{array}{r} 6\frac{3}{8} \\ -1\frac{1}{4} \\ \hline \end{array}$

17. $\begin{array}{r} 8\frac{2}{5} \\ -7\frac{9}{10} \\ \hline \end{array}$

18. $\begin{array}{r} 9\frac{1}{6} \\ -1\frac{4}{5} \\ \hline \end{array}$

19. $\begin{array}{r} 7\frac{2}{5} \\ -5\frac{3}{5} \\ \hline \end{array}$

20. $\begin{array}{r} 15\frac{3}{4} \\ -11\frac{1}{3} \\ \hline \end{array}$

3.5   Adding Decimals (DOK 1)

Example 6:   Find $0.7 + 3.2 + 78.19$.

Step 1:   When you add decimals, first arrange the numbers in columns with the decimal points under each other.

$$\begin{array}{r} 0.7 \\ 3.2 \\ +\ \ 78.19 \\ \hline \end{array}$$

Step 2:
Step 3:   Add 0's here to keep your columns straight. Start at the right and add each column. Remember to carry when necessary. Bring down the decimal point.

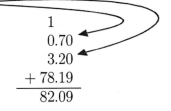

$$\begin{array}{r} 1 \\ 0.70 \\ 3.20 \\ +\ 78.19 \\ \hline 82.09 \end{array}$$

Add. Be sure to write the decimal point in your answer. (DOK 1)

1. $7.8 + 3.02 + 1.2$

2. $9.3 + 4.17 + 0.175$

3. $11.5 + 9.5 + 0.011$

4. $\$3.25 + \$8.99$

5. $\$0.31 + \$4.78$

6. $10.1 + 6.05 + 9.025$

7. $1.7 + 6.32 + 6.935$

8. $\$7.15 + \$7.15$

9. $9.35 + 0.001 + 1.2$

10. $\$19.95 + \$45.78$

11. $3.09 + 17.1 + 0.007$

12. $8.4 + 8.51 + 3.101$

13. $1.004 + 5.9 + 0.68$

14. $\$1.19 + \$89.50$

15. $\$4.23 + \$0.98$

16. $13.6 + 9.21 + 7$

17. $2.449 + 1.01$

18. $10.02 + 41.0997$

19. $\$103.29 + \$7.68$

20. $144.02 + 7.934 + 2.5$

21. $3.27 + 8.0054 + 1.1$

3.6   Subtracting Decimals (DOK 1)

Example 7:   Find $15.2 - 1.019$.

Step 1:   When you subtract decimals, arrange the numbers in columns with the decimal points under each other.

$$\begin{array}{r} 15.2 \\ -\ 1.019 \\ \hline \end{array}$$

Step 2:   You must fill in the empty places with 0's so that both numbers have the same number of digits after the decimal point.

$$\begin{array}{r} 15.200 \\ -\ 1.019 \\ \hline \end{array}$$

Step 3:   Start at the right and subtract each column. Remember to borrow when necessary.

$$\begin{array}{r} 19_1 \\ 15.\cancel{2}00 \\ -\ 1.019 \\ \hline 14.181 \end{array}$$

Subtract. Be sure to write the decimal point in your answer. (DOK 1)

1. $6.32 - 1.9$

2. $35.247 - 5.99$

3. $84.23 - 10.719$

4. $\$107.92 - \$22.46$

5. $\$50.00 - \$49.13$

6. $47.21 - 17.178$

7. $301.27 - 297.354$

8. $\$78.17 - \$32.45$

9. $\$119.10 - \$5.24$

10. $\$12.07 - \$9.02$

11. $9.934 - 8.013$

12. $17.8 - 6.41$

13. $6.2 - 4.97$

14. $\$25.49 - \$0.78$

15. $136.209 - 14.1$

16. $19.7 - 16.2$

17. $37.091 - 30.07$

18. $8.25 - 7.001$

19. $141.9 - 17.865$

20. $54.7821 - 10.09$

21. $12.42 - 3.235$

3.7 Representing Addition and Subtraction on a Number Line (DOK 2)

A number line can show important relationships between numbers. When adding two numbers, $a+b$, the result is a distance b from a and lies either to the right or left of a, depending on the sign of b. The distance between two numbers on a number line is the absolute value of their difference.

Example 8: Using the number line below, find the distance between $1\frac{1}{3}$ and $5\frac{2}{3}$

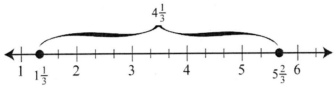

$$5\frac{2}{3} - 1\frac{1}{3} = 4\frac{1}{3}$$

Note that the distance between $5\frac{2}{3}$ and $1\frac{1}{3}$ is the absolute value of their difference $\left(5\frac{2}{3} - 1\frac{1}{3}\right)$. Also $5\frac{2}{3} - 1\frac{1}{3}$ can be written as $5\frac{2}{3} + \left(-1\frac{1}{3}\right)$. Remember, the "distance between" and subtraction are different types of questions.

Solve the following problems. (DOK 2)

1. Find $2.4 + 3.8$ using the number line below.

2. Add 4.3 and its opposite using the number line below.

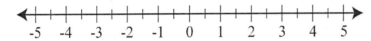

3. Solve $3.5 - 7.2$ using the number line below.

4. Solve $4\frac{1}{6} - 5\frac{3}{6}$ using the number line below.

5. What is the distance between $12\frac{4}{5}$ and $17\frac{2}{5}$? Use the number line below.

6. What is the distance between 20.3 and 27.4? Use the number line below.

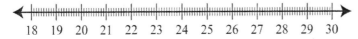

3.8   Real World Addition and Subtraction Problems (DOK 2)

Example 9:   The temperature on a February morning started at $-17°$F. By noon, the temperature had risen $8°$, and up an additional $12°$ by 3 p.m. By 10 p.m., the temperature fell $15°$. What is the temperature at 10 p.m.?

Step 1:   At noon, the temperature is $-17° + 8° = -9°$.

Step 2:   At 3 p.m., the temperature is $-9° + 12° = 3°$.

Step 3:   At 10 p.m., the temperature is $3° - 15° = -12°$.

Answer: The temperature is $-12°$ F at 10 p.m..

Solve the following problems. (DOK 2)

1. Henri owns a cupcake shop that also takes special orders. Henri starts the day baking and frosting 120 dozen cupcakes before the store opens. On Wednesday, he received a special order for 20 dozen assorted cupcakes to be delivered that same afternoon. He also sold 98 dozen cupcakes that day. How many dozen cupcakes does Henri have left at the end of the day?

2. Whitney is a tap dancer and wants to do a dance when she descends fifteen stairs to get to the front door of her house. She starts at the top of the stairs. She goes down 5 stairs and back up two stairs. Then she goes down 7 stairs and back up three stairs. How many stairs does she have left to descend to get to the front door?

3. Sara works for a movie theater and sells candy by the pound. Her first customer bought $2\frac{1}{8}$ pounds of candy, the second bought $\frac{4}{5}$ of a pound, and the third bought $\frac{1}{2}$ pound. How many pounds did she sell to the first three customers?

4. A chemical plant takes in $6\frac{1}{3}$ million gallons of water from a local river and discharges $2\frac{7}{12}$ million back into the river. How much water does not go back into the river?

5. The company water cooler started with $5\frac{3}{8}$ gallons of water. Employees drank $2\frac{1}{4}$ gallons. How many gallons were left in the cooler?

6. Fleta owns a candy store. On Monday, she sold 7.2 pounds of chocolate, 14.7 pounds of jelly beans, 2.5 pounds of sour snaps, and 9.03 pounds of yogurt-covered raisins. How many pounds of candy did she sell total?

3.9   Going Deeper into Adding and Subtracting Rational Numbers (DOK 3)

Solve the multi-step problems below. Express your answers in <u>decimal</u> form. Show your work for each step. (DOK 3)

1. Mr. Gordon bought $2\frac{1}{2}$ pounds of apples, 1.25 pounds of pears, and $3\frac{3}{4}$ pounds of oranges. He gives 1.1 pounds of apples, $\frac{3}{4}$ pound of pears, and 1.5 pounds of oranges to his sister on the way home. How many pounds of fruit did Mr. Gordon take to his own home?

2. The day starts at $-3°$ F. By 10:00 a.m., it is $17.5°$ warmer. By 3:00 p.m. it is $15°$ warmer than it was at 10:00 a.m. At 6:00 p.m. it is $7.5°$ cooler than it was at 3:00 p.m. Finally, at 10:00 p.m. it is $12.5°$ cooler than it was at 6:00 p.m. What is the temperature at 10:00 p.m.?

3. Alan has $115.78 in his savings account. He earns $15.00 cleaning out the garage and puts $7.50 into savings. He withdraws $35.90 to buy a video game. He receives $20.00 from his grandfather for his birthday and deposits $15.00 into his savings account. What is Alan's balance now?

4. Sandy bought $12\frac{7}{8}$ lb of peaches to make peach jam. She and her mother eat 0.75 of a pound of the peaches before making the jam. They use $9\frac{1}{8}$ pounds of the peaches to make the jam. How many pounds of peaches are left over?

5. Jerry knows he watches too much TV. He decides to track the amount of time he watches TV for one week. The results are in the table below.

Mon	Tues	Wed	Thur	Fri	Sat	Sun
4.25	$3\frac{1}{2}$	3.75	$4\frac{1}{2}$	$5\frac{3}{4}$	12.25	11.5

Jerry decides to withdraw from his TV viewing by watching $2\frac{1}{2}$ hours less each day. Fill in the chart below showing how many hours Jerry will allow himself to watch TV next week.

Mon	Tues	Wed	Thur	Fri	Sat	Sun

6. Trudi's dog squirms too much on the bathroom scale to get weighed. Also, her dog won't let go of his dog toy in his mouth. Trudi decides to find out his weight by picking up her dog with his dog toy hanging from his mouth and stepping on the scale. Together, they weigh 152.4 pounds. She steps off the scale, puts down the dog and steps back on the scale by herself. Alone, she weighs 128.8 pounds. When her dog dropped his toy, she picked it up and weighed it on the counter top scale in the kitchen. The toy weighs 0.35 pounds. How much does the dog weigh by himself?

Chapter 3 Review

Simplify the following improper fractions. (DOK 1)

1. $\dfrac{14}{8} = $ _____   2. $\dfrac{35}{4} = $ _____   3. $\dfrac{16}{5} = $ _____   4. $\dfrac{9}{4} = $ _____   5. $\dfrac{62}{3} = $ _____

Solve the following fraction problems using addition and subtraction. (DOK 1)

6. $4\frac{1}{4}$ 7. $8\frac{2}{5}$ 8. $6\frac{2}{9}$ 9. $7\frac{3}{7}$ 10. $12\frac{3}{8}$ 11. $9\frac{1}{6}$

   $+1\frac{1}{2}$ $-3\frac{3}{4}$ $+4\frac{1}{3}$ $+2\frac{1}{14}$ $+5\frac{1}{4}$ $-\frac{2}{3}$

Solve the following decimal problems using addition and subtraction. (DOK 1)

12. $0.12 - 0.09$ 13. $4.15 + 6.17$ 14. $583.62 - 472.07$ 15. $14.238 + 1.14 + 0.32$

Carefully read each of the following problems and answer. (DOK 2)

16. Alicia and her mom were on the way to the shopping mall when they were pulled over by a police officer. The officer said Alicia's mom was going too slow at -8 miles under the minimum speed of 45 miles per hour. How fast was Alicia's mom going?

17. Candace weighs 97 pounds. Her younger brother, Andrew weighs 82 pounds. How many pounds different is their weight?

18. At 8:00 AM, the temperature was $46°$ F. By noon the temperature rose 15 degrees. A blizzard comes through and drops the temperature 26 degrees by 4:00 PM. What is the temperature at 4:00 PM?

Solve the multi-step problems below. Show your work for each step. (DOK 3)

19. It's the morning of the school carnival and 16 cakes are on the table, so far. Cakes are being sold whole and by the slice. Families arrive and buy a total of $12\frac{3}{4}$ cakes by noon. Fourteen more cakes are dropped off for sale. Another $17\frac{1}{8}$ cakes are sold by the end of the day. How much cake is leftover?

20. The day starts at $72°$ F. By noon, it is $23°$ warmer. At 3:00 p.m. it is $14.5°$ hotter than at noon. At 6:00 p.m. it is $7.5°$ cooler than at 3:00 p.m. At dusk it is $11.5°$ cooler than it was at 6:00 p.m. What is the temperature at dusk?

21. Find the distance between -3.2 and 2.6 using the number line below.

22. Solve $4\frac{9}{10} - 1\frac{1}{5}$ using the number line below.

Chapter 3 Test

1 Simplify: $\dfrac{52}{3}$

  A $17\frac{1}{3}$

  B 18

  C $18\frac{1}{3}$

  D 19

(DOK 1)

2 $28\frac{1}{4} + 12\frac{5}{16} =$

  A $40\frac{1}{4}$

  B $40\frac{6}{16}$

  C $40\frac{9}{16}$

  D $41\frac{1}{4}$

(DOK 1)

3 $977.62 - 128.41 =$

  A 841.49

  B 845.21

  C 847.49

  D 849.21

(DOK 1)

4 Simplify: $\dfrac{17}{4}$

  A 4

  B $4\frac{1}{4}$

  C $4\frac{3}{4}$

  D $5\frac{1}{4}$

(DOK 1)

5 $22\frac{1}{15} - 4\frac{3}{5} =$

  A $18\frac{7}{15}$

  B $18\frac{2}{15}$

  C $17\frac{2}{15}$

  D $17\frac{7}{15}$

(DOK 2)

6 $8.7 + 55.33 =$

  A 63.4

  B 63.43

  C 64.03

  D 64.4

(DOK 1)

7 $-22.1 + (-16.8) =$

  A 38.9

  B 5.3

  C -38.9

  D -39.8

(DOK 1)

8 One pizza was cut into 12 pieces. Eddie ate $\frac{1}{12}$ of the pizza, Melanie ate $\frac{1}{12}$ of the pizza, Mom ate $\frac{1}{6}$ of the pizza, Dad ate $\frac{1}{4}$ of the pizza, and Uncle Albert ate $\frac{1}{6}$ of the pizza. How much, if any, pizza was left over?

  A $\frac{1}{12}$

  B $\frac{1}{9}$

  C $\frac{1}{6}$

  D $\frac{1}{4}$

(DOK 2)

9 Find the distance between 4.1 and 2.9, using the number line below.

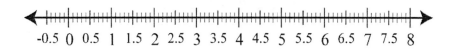

A 1.2

B 7.0

C −1.2

D −7.0

(DOK 2)

10 Mr. Jackson bought a 12 pound container of nails. He used 3.45 pounds on an addition for his deck. He used $1\frac{1}{4}$ pounds to make a new play fort for his children. He already had $2\frac{1}{2}$ pounds left over from another container. How many pounds of nails does Mr. Jackson have now?

A 7.3 lb

B 9.8 lb

C 8.8 lb

D 8.3 lb

(DOK 3)

11 Two cousins that live in different places were comparing temperatures in their home towns. The chart below shows their results for a day in May.

Cousin	Dawn	Noon	3:00 p.m.	Dusk
Jeremy	31°	16° warmer than dawn	12° warmer than noon	18° cooler than 3:00 p.m.
Joshua	54°	18° warmer than dawn	22° warmer than noon	25° cooler than 3:00 p.m.

Find the difference between the highest temperature of Joshua's data and the lowest temperature of Jeremy's data.

A 53°

B 41°

C 63°

D 88°

(DOK 3)

Chapter 4
Multiplying and Dividing Rational Numbers

This chapter covers the following CC Grade 7 standards:

	Content Standards
Expressions and Equations	7.NS.2, 7.NS.3

4.1   Multiplying Fractions (DOK 1)

Example 1:   Multiply: $4\frac{3}{8} \times \frac{8}{10}$

📱 Refer to the "About AR" on page xi!

Step 1:   Change the mixed numbers in the problem to improper fractions. $\dfrac{35}{8} \times \dfrac{8}{10}$

Step 2:   When multiplying fractions, you can cancel and simplify terms that have a common factor.

The 8 in the first fraction will cancel with the 8 in the second fraction.

$$\dfrac{35}{\cancel{8}_1} \times \dfrac{\cancel{8}^1}{10}$$

The terms 35 and 10 are both divisible by 5, so

35 simplifies to 7, and 10 simplifies to 2. $\dfrac{\overset{7}{\cancel{35}}}{1} \times \dfrac{1}{\underset{2}{\cancel{10}}}$

Step 3:   Multiply the simplified fractions. $\dfrac{7}{1} \times \dfrac{1}{2} = \dfrac{7}{2} = 3\frac{1}{2}$

Multiply and simplify your answers. (DOK 1)

1. $4\frac{1}{2} \times 2\frac{2}{3}$

9. $2\frac{1}{5} \times 1\frac{7}{8}$

17. $4\frac{1}{8} \times 4$

2. $\frac{4}{5} \times 6\frac{1}{8}$

10. $\frac{4}{7} \times 6\frac{3}{4}$

18. $\frac{2}{5} \times 3\frac{1}{4}$

3. $3\frac{1}{9} \times 2\frac{2}{7}$

11. $2\frac{4}{5} \times 2\frac{4}{7}$

19. $5\frac{1}{9} \times 7\frac{1}{2}$

4. $\frac{6}{7} \times 1\frac{2}{9}$

12. $3\frac{1}{3} \times \frac{5}{6}$

20. $6\frac{4}{5} \times 2\frac{1}{2}$

5. $3 \times 2\frac{1}{4}$

13. $1\frac{1}{14} \times 7\frac{3}{5}$

21. $4\frac{1}{3} \times 1\frac{1}{5}$

6. $1\frac{1}{8} \times 1\frac{3}{4}$

14. $2\frac{3}{4} \times 1\frac{1}{2}$

22. $2\frac{1}{5} \times 4\frac{2}{7}$

7. $4 \times \frac{6}{13}$

15. $2\frac{1}{2} \times 3\frac{1}{4}$

23. $1\frac{3}{4} \times 1\frac{7}{9}$

8. $5\frac{3}{5} \times 3\frac{1}{7}$

16. $6\frac{2}{3} \times \frac{9}{10}$

24. $3\frac{2}{5} \times 1\frac{1}{4}$

4.2   Dividing Fractions (DOK 1)

Example 2:   Divide: $1\frac{3}{4} \div 2\frac{5}{8}$

Refer to the "About AR" on page xi!

Step 1:   Change the mixed numbers in the problem to improper fractions. $\frac{7}{4} \div \frac{21}{8}$.

Step 2:   Take the reciprocal (flip it upside down) of the second fraction and multiply.
$\frac{7}{4} \times \frac{8}{21}$

Step 3:   Cancel where possible and multiply. $\overset{1}{\frac{7}{4}} \times \overset{2}{\frac{8}{21}} = \frac{2}{3}$

Divide and simplify answers to lowest terms. (DOK 1)

1. $8\frac{1}{4} \div \frac{3}{8}$

2. $4 \div 1\frac{1}{7}$

3. $2\frac{1}{2} \div \frac{5}{8}$

4. $7\frac{5}{9} \div 7\frac{4}{9}$

5. $6 \div \frac{3}{4}$

6. $7\frac{1}{2} \div 4\frac{1}{8}$

7. $2\frac{1}{5} \div 3\frac{2}{5}$

8. $5\frac{1}{3} \div 2\frac{2}{9}$

9. $\frac{3}{4} \div 2$

10. $5\frac{5}{9} \div 1\frac{1}{3}$

11. $2\frac{1}{4} \div 7\frac{1}{4}$

12. $9\frac{1}{5} \div 4\frac{3}{5}$

13. $10\frac{1}{12} \div 5\frac{1}{2}$

14. $1\frac{4}{5} \div 6\frac{3}{10}$

15. $4\frac{1}{2} \div 2$

16. $3\frac{1}{4} \div \frac{7}{12}$

17. $1\frac{1}{2} \div \frac{3}{4}$

18. $11\frac{2}{3} \div 4\frac{1}{6}$

19. $7\frac{2}{5} \div \frac{3}{5}$

20. $6\frac{1}{8} \div 3\frac{5}{8}$

21. $2\frac{1}{2} \div 1\frac{3}{8}$

22. $6\frac{2}{9} \div 2\frac{1}{3}$

23. $3\frac{1}{2} \div 2\frac{1}{4}$

24. $4\frac{4}{7} \div \frac{8}{14}$

4.3   Multiplication of Decimals (DOK 1)

Example 3:   Find 56.2×0.17.

Step 1:   Set up the problem as if you were multiplying whole numbers.

$$\begin{array}{r} 56.2 \\ \times\, 0.17 \\ \hline \end{array}$$

Step 2:   Multiply as if you were multiplying whole numbers.

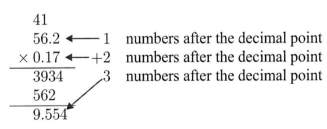

41
56.2 ←— 1 numbers after the decimal point
× 0.17 ←—+2 numbers after the decimal point
3934 ↘ 3 numbers after the decimal point
562
9.554

Step 3:   Count how many numbers are after the decimal points in the problem. In this problem, 2, 1, and 7 come after decimal points, so the answer must also have three numbers after the decimal point.

Multiply. (DOK 1)

1. 12.2×4.9	5. 37.2×4.6	9. 14.75×3.3	13. 17.69×2.3
2. 6.5×7.83	6. 0.074×18	10. 7.92×0.11	14. 0.56×4.1
3. 21.07×1.5	7. 0.47×0.816	11. 9.8×0.13	15. 6.44×9.01
4. 0.057×1.7	8. 11.91×14.1	12. 4.5×1.57	16. 2.37×0.49

4.4   More Multiplying Decimals (DOK 1)

Example 4:   Find 0.003×0.723.

Step 1:   Multiply as you would whole numbers.

$$
\begin{array}{r}
0.003 \quad \longleftarrow 3 \quad \text{numbers after the decimal point} \\
\times\ 0.723 \quad \longleftarrow +3 \quad \text{numbers after the decimal point} \\
\hline
0.002169 \quad \longleftarrow 6 \quad \text{numbers after the decimal point}
\end{array}
$$

Step 2:   Count how many numbers are behind decimal points in the problem. In this case, six numbers come after decimal points in the problem, so there must be six numbers after the decimal point in the answer. In this problem, 0's needed to be written in the answer in front of the 2, so there would be 6 numbers after the decimal point.

Multiply. Write in zeros as needed. Round dollar figures to the nearest penny. (DOK 1)

1. 4.7×0.003	5. 6.37×2.15	9. 7.81×1.53	13. 0.004×0.155
2. 12.3×1.9	6. $\$5.54 \times 0.07$	10. $\$13.00 \times 0.012$	14. 4.6×0.918
3. 1.9×0.17	7. 6.44×0.02	11. 7.89×9.1	15. 3.15×3.015
4. 45.6×0.113	8. $\$9.84 \times 0.02$	12. 4.31×9.7	16. 0.022×4.41

4.5   Division of Decimals By Whole Numbers (DOK 1)

Example 5:   Find $52.26 \div 6$.

Step 1:   Copy the problem as you would for whole numbers. Copy the decimal point directly above in the place for the answer.

$$6\overline{)52.26}$$

Step 2:   Divide the same way as you would with whole numbers.

$$
\begin{array}{r}
8.71 \\
6\overline{)52.26} \\
-48 \\
\hline
4\ 2 \\
-4\ 2 \\
\hline
6 \\
-6 \\
\hline
0
\end{array}
$$

Divide. Remember to copy the decimal point directly above the place for the answer. (DOK 1)

1. $47.93 \div 2$
2. $82.14 \div 5$
3. $191.06 \div 2$
4. $94.6 \div 12$

5. $13.221 \div 3$
6. $44.97 \div 6$
7. $12.99 \div 3$
8. $37.7 \div 15$

9. $189.04 \div 12$
10. $109.68 \div 20$
11. $10.77 \div 4$
12. $18.75 \div 5$

13. $368.5 \div 5$
14. $49.88 \div 25$
15. $146.2 \div 12$
16. $98.9 \div 16$

4.6   Division of Decimals by Decimals (DOK 1)

Example 6: $374.5 \div 0.07$

Step 1:   Copy the problem as you would for whole numbers.

$$\overset{\text{Divisor}}{\swarrow}$$

$0.07\overline{)374.5}$ ◄——— **Dividend**

Step 2:   You cannot divide by a decimal number. You must move the decimal point in the divisor 2 places to the right to make it a whole number. The decimal point in the dividend must also move to the right the same number of places. Notice that in this example, you must add a 0 to the dividend.

$0.07.\overline{)374.50.}$

Step 3:   The problem now becomes $37450 \div 7$. Copy the decimal point from the dividend straight above in the place for the answer.

```
          5350.
007.)37450.
    - 35
      24
    - 21
       35
     - 35
       00
```

Divide. Remember to move the decimal points. (DOK 1)

1. $0.676 \div 0.013$

2. $70.32 \div 0.08$

3. $\$54.60 \div 0.84$

4. $\$10.35 \div 0.45$

5. $18.46 \div 1.3$

6. $14.6 \div 0.002$

7. $\$125.25 \div 0.75$

8. $\$33.00 \div 1.65$

9. $154.08 \div 1.8$

10. $0.4374 \div 0.003$

11. $292.9 \div 0.29$

12. $6.375 \div 0.3$

13. $4.8 \div 0.08$

14. $1.2 \div 0.024$

15. $15.725 \div 3.7$

16. $\$167.50 \div 0.25$

4.7   Real World Multiplication and Division Problems (DOK 2)

Example 7:   Megan can buy 6 boxes of pita chips for $2.22, 8 boxes of pita chips for $3.76, or 5 boxes of pita chips for $3.00. Which option offers the least expensive unit price for one box of pita chips?

    Step 1: $2.22 \div 6 = \$0.37$ per box

    Step 2: $3.76 \div 8 = \$0.47$ per box

    Step 3: $3.00 \div 5 = \$0.60$ per box

    Answer: The first option (6 boxes for $2.22) offers the least expensive unit price for one box of pita chips.

Solve the following problems. (DOK 2)

1. Sharikka and her friends were playing a game that included a pile of cards telling them to move forward or backwards a certain number of spaces. Sharikka sorted the cards by the amount written on each card. She found six cards that said "move -3 spaces." If one player received all six cards that said "move -3 spaces," how many spaces would that player move?

2. A plane is flying at 2,000 feet and starts to descend 50 feet every 10 seconds. What will the altitude be of the plane after one minute?

3. George has $5.00 to spend on candy. If each candy bar costs $0.71, how many bars can he buy?

4. Juan was competing in a 1,000 meter race. He had to pull out of the race after running $\frac{7}{8}$ of it. How many meters did he run?

5. Tad needs to measure where the free throw line should be in front of his basketball goal. He knows his feet are $\frac{3}{4}$ feet long and the free-throw line should be 15 feet from the backboard. How many toe-to-toe steps does Tad need to take to mark off 15 feet?

6. Beth has a bread machine that makes a loaf of bread that weighs $2\frac{1}{2}$ pounds. If she makes a loaf of bread for each of her three sisters, how many pounds of bread will she make?

4.8   Going Deeper into Multiplying and Dividing Rational Numbers (DOK 3)

Solve the multi-step problems below. Show your work for each step. (DOK 3)

1. The purchasing department for a factory that makes stuffed animals is trying to estimate the number of $\frac{3}{4}$ inch eyes needed for medium size teddy bears for one year. The eyes come in boxes of 600 eyes per box. The purchasing department has received estimates from the sales department of selling an average of 28, 500 medium teddy bears per month. How many boxes of $\frac{3}{4}$ inch eyes will the purchasing department buy to keep production going for one year? (Remember each of the teddy bears have 2 eyes.)

2. There are 52 crackers in each 10 oz box. A restaurant buys 48 boxes of crackers to serve with their soups. Each person will receive 4 crackers. Assuming the customers do not ask for extra crackers, how many customers can the restaurant serve with 48 boxes of crackers?

3. The McTavish family is taking a cross country trip by minivan from Miami, FL to Seattle, WA for a total of 3, 315 miles. Mr. McTavish knows that he gets an average of 19.4 miles to the gallon and as of this writing, gas is an average of $3.34 per gallon, nationwide. Not including little side trips on the way to Seattle, how much money will Mr. McTavish have to pay for gas to get his family to Seattle? (Round the number of gallons to the nearest whole gallon before calculating the total cost.)

4. Five brothers decide to go together to buy a new video game system. The total price is $456.31, including taxes.

Part 1: If they five brothers equally share the price of the system, how much will each brother pay? Round your answer to the nearest penny.

Part 2: Suppose the grandparents decide to put up half the share of the twin brothers for their birthday. The five brothers decide to redistribute the remaining cost of the system among them. How much would each brother pay now? Round your answer to the nearest penny.

5. Carlton Middle School's cafeteria is going to serve 2 cookies to each of 1, 230 students that take hot lunch. The recipe they use makes 615 cookies. Some of the ingredients are in the table below.

One Recipe: Makes 615 Cookies	
Butter	6.4 lb
White Sugar	18.75 cups
Brown Sugar	15.25 cups
Eggs	51

Part 1: How many batches will the cafeteria cooks have to make for each student taking hot lunch to receive 2 cookies?

Part 2: Multiply each of the ingredients in the table above by your answer in part 1. This is the amount of each ingredient needed to make all the cookies.

Chapter 4 Review

Solve the following fraction problems using multiplication or division. (DOK 1)

1. $6 \times 8\frac{1}{2} =$
4. $5\frac{1}{2} \div \frac{1}{4} =$
3. $4\frac{4}{6} \times 1\frac{5}{7}$
6. $6\frac{6}{7} \div 2\frac{2}{3}$

2. $18 \div \dfrac{1}{2} =$
1. $1\frac{1}{3} \times 3\frac{1}{2}$
4. $\frac{2}{3} \times \frac{5}{6}$
7. $3\frac{5}{6} \div 11\frac{1}{2}$

3. $2\frac{1}{2} \times 3\frac{1}{3} =$
2. $5\frac{3}{7} \times \frac{7}{8}$
5. $\frac{1}{2} \div \frac{4}{5}$
8. $1\frac{1}{3} \div 3\frac{1}{5}$

Solve the following decimal problems using multiplication or division. (DOK 1)

13. 8.32×0.44
16. 0.52×11.24
19. 30.7×0.0041

14. $9 \div 1.5$
17. 4.58×0.025
20. $17.28 \div 0.054$

15. $36.6 \div 0.6$
18. 0.879×1.7
21. $174.66 \div 1.23$

Carefully read each of the following problems and answer. (DOK 2)

22. The Acme Aluminum Can Company sells its aluminum cans by the pound. A case of cans weighs 14 pounds. If the Green Bean Company orders 18 cases of aluminum cans, how many pounds will they receive?

23. Terri is packing packages of cookies into a bigger box. A big box will hold 5 packages of cookies. If she has 37 packages of cookies, how many big boxes will she need?

24. Jane is sewing buttons on sweaters. She has 3 new sweaters and she is sewing 3 buttons on each sweater. If buttons come in packs of 5, how many packs will she need?

Solve the multi-step problems below. Show your work for each step. (DOK 3)

25. A restaurant buys eggs in cases of 15 dozen eggs per case. The owner of the restaurant specializes in omelets. He orders 3 cases of eggs per day just for omelets. If each omelet contains 4 eggs, how many omelets can be made from 3 cases of eggs?

26. There are 16 tablespoons in one cup. There are 3 teaspoons in one tablespoon. Lisa is making cookies and realizes too late that she doesn't have any sugar left in the canister. She does have a large box of sugar packets, each container $\frac{1}{2}$ teaspoon of sugar. If she needs $\frac{3}{4}$ cup of sugar for her recipe, how many packets of sugar will she use?

Chapter 4 Test

1 $7\frac{1}{3} \times 2\frac{1}{2} =$

A $14\frac{1}{6}$

B $14\frac{1}{3}$

C $16\frac{1}{3}$

D $18\frac{1}{3}$

(DOK 1)

2 $27 \div 1\frac{1}{4} =$

A 27

B $21\frac{3}{5}$

C 21

D $13\frac{1}{7}$

(DOK 1)

3 $11.73 \div 5.1 =$

A 2.1

B 2.3

C 2.35

D 2.4

(DOK 1)

4 $8.3 \times 0.2 =$

A 166

B 160.6

C 16.6

D 1.66

(DOK 1)

5 At the store, a 12 ounce can of soda costs $0.75. What is the cost of one ounce of this soda?

A $0.625

B $6.25

C $0.0625

D $0.00625

(DOK 2)

6 Sara and Sharelle are baking cherry pies for a community supper. If each pie takes 4 cups of cherries, and a container of cherries holds 8 cups, how many containers of cherries will the girls need to make 10 pies?

A 4

B 3

C 5

D 9

(DOK 2)

7 The senior class of 361 students is going on a field trip. If each bus holds 40 people and there are 20 chaperones going, how many buses will they need?

A 9

B 10

C 11

D 8

(DOK 2)

8 There are 12 small hair clips in each package of hair clips. A large chain of stores places an order for 7,000 packages to be evenly disbursed among 40 different stores. How many hair clips will each store receive?

A 2,100

B 2,200

C 21,000

D 22,000

(DOK 3)

Chapter 5
Percents

This chapter covers the following CC Grade 7 standard:

	Content Standard
Ratios and Proportional Relationships	7.RP.3

5.1   Changing Percents to Decimals and Decimals to Percents (DOK 1)

To change a **percent** to a **decimal**, move the **decimal** point two places to the left, and drop the **percent** sign. If there is no decimal point shown, it is understood to be after the number and before the percent sign. Sometimes you will need to add a "0". (See 8% below.)

Example 1: $23\% = 0.23$ $8\% = 0.08$ $100\% = 1$ $409\% = 4.09$

(decimal point)

Change the following percents to decimal numbers. (DOK 1)

1.	35%	5.	100%	9.	70%	13.	644%	17.	22%
2.	98%	6.	62%	10.	33%	14.	12%	18.	7%
3.	9%	7.	432%	11.	800%	15.	90%	19.	500%
4.	10%	8.	19%	12.	2%	16.	14%	20.	87%

To change a decimal to a percent, move the decimal two places to the right, and add a percent sign. You may need to add a "0". (See 0.2 below.)

Example 2: $0.24 = 24\%$ $0.03 = 3\%$ $0.2 = 20\%$ $0.445 = 44.5\%$ $2.37 = 237\%$

Change the following decimal numbers to percents. (DOK 1)

21.	0.26	25.	1.11	29.	8.52	33.	5.55	37.	3.26
22.	0.84	26.	0.214	30.	2.33	34.	4.04	38.	0.75
23.	6.52	27.	1.8	31.	0.77	35.	0.002	39.	0.1
24.	0.99	28.	0.003	32.	0.08	36.	0.67	40.	9.44

5.2   Changing Percents to Fractions and Fractions to Percents (DOK 1)

Example 3:   Change 135% to a fraction.

Step 1:   Copy the number without the percent sign. 135 is the numerator (the top number) of the fraction.

Step 2:   Anytime you change a percent to a decimal, the denominator of the fraction is 100.

$$135\% = \frac{135}{100}$$

Step 3:   Simplify the fraction. $\frac{135}{100} = 1\frac{7}{20}$

Change the following percents to fractions and reduce. (DOK 1)

1.	42%	5.	200%	9.	25%	13.	500%	17.	157%
2.	17%	6.	84%	10.	83%	14.	9%	18.	333%
3.	180%	7.	110%	11.	350%	15.	61%	19.	29%
4.	3%	8.	37%	12.	10%	16.	88%	20.	50%

Example 4:   Change $2\frac{3}{4}$ to a percent.

Step 1:   Change the mixed number $2\frac{3}{4}$ to an improper fraction. $2\frac{3}{4} = \frac{11}{4}$

Step 2:   Divide the numerator by the denominator. $11 \div 4$

$$\begin{array}{r} 2.75 \\ 4\overline{)11.00} \\ -8 \\ \hline 3\,0 \\ -2\,8 \\ \hline 20 \\ -20 \\ \hline 0 \end{array}$$

Step 3:   Change the decimal answer, 2.75, to a percent by moving the decimal point 2 places to the right.

$$2\frac{3}{4} = 2.75 = 275\%$$

Change the following fractions and mixed numbers to percents. (DOK 1)

1. $\frac{1}{2}$ 5. $\frac{8}{32}$ 9. $4\frac{4}{5}$ 13. $\frac{99}{100}$ 17. $7\frac{1}{4}$

2. $\frac{9}{10}$ 6. $3\frac{7}{8}$ 10. $\frac{1}{4}$ 14. $\frac{45}{100}$ 18. $\frac{3}{8}$

3. $2\frac{1}{2}$ 7. $9\frac{3}{4}$ 11. $5\frac{3}{8}$ 15. $3\frac{1}{2}$ 19. $\frac{1}{5}$

4. $3\frac{1}{4}$ 8. $\frac{1}{16}$ 12. $12\frac{1}{2}$ 16. $\frac{7}{8}$ 20. $6\frac{1}{4}$

5.3   Percent Word Problems (DOK 2)

Example 5:   Wayne has spent 40% of his allowance on paper for his computer printer. If Wayne spent exactly $3.60 including tax on the paper, how much is his allowance, and how much does he have left to spend?

Step 1:   Turn the percentage from the problem into a decimal. $40\% = 0.4$

Step 2:   Divide the dollar amount by the percent. $\$3.60 \div 0.4 = \$9$

Step 3:   Wayne's allowance is $9.00 and he has $\$9.00 - \$3.60 = \$5.40$ left over.

Carefully read the word problems below and solve. (DOK 2)

1. Emily is allowed to spend 60% of her allowance. The remaining 40% she deposits into her savings account. Emily deposited $2.00 into her savings account this week. What amount does Emily get each week for her allowance?

2. Dashon walks 30 miles a week for exercise. He has completed 40% of his weekly goal. How many more miles does Dashon have to walk to meet his goal?

3. Gretchen wants to make a quilt for her bed. She has completed 60% of the quilt. When completed, the quilt will be made up of 90 pieces. How many pieces does Gretchen have left to sew?

4. Hank is buying a shirt that originally cost $25.00. It was on sale two weeks ago for 10% off. Today, it is an additional 20% off the **original** price. How much is the shirt now?

5. Alicia is walking around the shopping mall with her mom to find the best price on a new blouse. One store has a blouse she wants for 25% off the original price of $40.00. Another store has an almost identical blouse for 30% off the original price of $50.00. Which blouse will cost less, the one priced 25% off or the one priced 30% off?

5.4   Finding the Percent of the Total (DOK 2)

Example 6:   800 people came to the high school football game. Sixty-five percent of the people came to cheer for the home team. How many people came to cheer for the home team?

Step 1:   Change 65% to a decimal. $65\% = 0.65$

Step 2:   Multiply the percent by the number of people who attended the game. $800 \times 0.65 = 520$

520 people came to cheer the home team.

Carefully read the problems below and solve. (DOK 2)

1. Lyla baked 200 cupcakes for the school bake sale. She sold 80% of them. How many did she sell?

2. Hill's Department Store had a one day sale on footballs just before the start of the fall football season. They started the day with 1,000 footballs, and sold 87 percent of the footballs. How many footballs did the store have left at the end of the day?

3. Fifteen percent of a bag of chocolate candies have a red coating on them. How many red pieces are in a bag of 60 candies?

4. Cary's dad gave her $20.00 to spend on shoes at the mall. Cary spent 85% of the money on one pair of shoes. How much money does Cary's dad get back from his $20.00?

5. The lunchroom at Prairie Middle School sold 90% of the burger meat they ordered for the week. If Mrs. Halgurst ordered 900 pounds of meat for a week's worth of burgers, how many pounds of meat did the lunchroom sell?

6. Thirty percent of sixty students polled said dogs were their favorite pets. How many of the sixty students preferred dogs?

7. Forty-five percent of the same sixty students polled said cats were their favorite pets. How many of the sixty students preferred cats?

8. Hilda works at a candy store and sold 80 pounds of peanut butter candies in one day. Computer totals showed that of the 120 customers that day, 40% bought peanut butter candies. How many customers purchased peanut butter candies that day?

9. Eighty-two percent of 300 boys polled said that they liked to play outdoors. How many boys liked to play outdoors?

5.5   Percent Increase or Decrease (DOK 2)

When finding the percent increase or decrease in a word problem, you must first identify the **original** number and then the **change.** The **original** number is identified by the value that takes place first in the word problem. It may or may not be mentioned first in the problem. You will need to study the word problem carefully to identify the values.

The **change** is the difference between the earlier value and the current value in the word problem. This can be figured out by simply subtracting the two numbers.

To find the percent increase or decrease divide the **change** by the **original** value.

$$\frac{\text{Change}}{\text{Original}}$$

Example 7:   The population of Wilkinsville was 3,200 one year and increased to 3,520 the next year. What is the percent **increase** of the population of Wilkinsville?

Step 1:   Find the difference in the two numbers. This equals the change in population.

$$3,520 - 3,200 = 320$$

Step 2:   Divide the difference by the original population. $320 \div 3,200 = 0.10$

Step 3:   Change the answer to a percent. $0.10 = 10\%$

The population increased by 10%.

Example 8:   Mr. Hunter had 25 students in his 4th period math class. The next day, he had only 20 students. What is the percent **decrease** in the number of students in Mr. Hunter's science class?

Step 1:   Find the difference in the two numbers. This equals the change in the number of students.

$$25 - 20 = 5$$

Step 2:   Divide the difference by the original number of students.

$$5 \div 25 = 0.20$$

Step 3:   Change the answer to a percent. $0.20 = 20\%$

The number of students in Mr. Hunter's class decreased by 20%.

Carefully read the problems below and solve. (DOK 2)

1. The middle school's baseball team, the Badgers, won 20 games last year. This year they won 22 games. What is the percent increase of games won from this year to last year?

2. Tony bought a printer on sale for $45.00. The original price was $60.00. What is the percent decrease in price on the printer?

3. If a loaf of bread was priced $1.50 on one day and $1.20 the next, what is the percent decrease of the price of the loaf of bread?

4. The Toys of Life store had 240 customers on October 24th. They had 264 customers on October 25th. What was the percent increase in customers from October 24th to October 25th?

5. Dr. Baler's veterinary practice is booming. Last year he had 1,500 appointments. This year he had 2,250 appointments and had to hire more help. What is the percent increase in appointments from this year to last year?

6. Lisa's little brother weighed 8 pounds when he was born. He now weighs 36 pounds. What is the percent increase in weight of the little boy?

7. John used to weigh 270 pounds and now weighs 216 pounds. What is the percent decrease in weight of John?

8. Rita Rose had 120 buttons in her button collection. She now has 150 buttons. What is the percent increase in the number of buttons in Rita Rose's collection?

9. Matthew had $50.00 in his special hiding place. He decides to spend $15.00 on his mom's birthday present. What percent decrease in money will this be for Matthew?

10. Farmer Jake harvested 2,500 pounds of carrots this fall. He sold them out of the back of his truck and brought home only 250 pounds after a full day of selling. What is the percent decrease in the number of carrots Farmer Jake has left?

5.6   Tips and Commissions (DOK 2)

In many businesses, sales people are paid on commission - a percent of the total sales they make. Waiters and waitresses are generally tipped a percentage of the total price of the meal and drinks their customers purchased.

Example 9:   Mrs. Hanson made 2% commission on a house she sold. The house sold for $150,000. How much was Mrs. Hanson's commission?

Step 1:   Change the percent of the commission to a decimal. $2\% = 0.02$

Step 2:   Multiply the percent commission by the total sale. $\$150,000 \times 0.02 = \$3,000$

Mrs. Hanson earned $3,000 commission on the sale of the house.

Carefully read the problems below and solve. (DOK 2)

1. Miss Sally waited on a large party at the Chesterfield Restaurant. The party's bill totaled $110.00. As the party was pleased with the service they received, they tipped Miss Sally 20%. What amount did Miss Sally receive as her tip?

2. Ed makes 6% commission on every case of widgets he sells. This week, Ed sold 16 cases for $800.00 each. How much did Ed make on commissions this week?

3. Maria makes $6.00 an hour plus 5% commission for telephone sales. Yesterday, Maria worked 8 hours and had sales of $600.00. How much did Maria make yesterday?

4. Alonzo earns 8% commission on vacuum cleaner sales. This past month, Alonzo sold 70 units at $150.00 each. How much did Alonzo make in commissions for the past month?

5. Rick was given a tip of 15% waiting on a table of customers who ordered $60.00 in food and drinks. How much did Rick earn in tips waiting on this table?

6. Acme Auto Parts gives all of their sales people an hourly wage and 3% commission on all the sales they ring up. Daniel rang up $3,200.00 last week. How much did Daniel earn in commissions for the week?

7. Tanner and Brad split the 6% commission they made selling a house for $120,000. How much did each of them receive?

8. Bridget's mom offered to pay her $5.00 to clean out the refrigerator and freezer, and if she got it done before dinner time, 6:00 p.m., she would receive a 20% bonus. Not one to pass up an opportunity, Bridget cleaned the refrigerator and freezer by 5:00 p.m. - an hour to spare! What is the total amount Bridget received from her mom?

5.7   Finding the Amount of Discount (DOK 2)

Prices are sometimes marked a percent off to sell remaining inventory in a hurry. The amount you save is called the **discount**.

Example 10:   A chair that sells for $159.00 is on sale for 30% off. How much can you save if you buy the chair on sale?

    Step 1:   Change the percent of the discount to a decimal. $30\% = 0.30$

    Step 2:   Multiply the original price of the chair by the discount.
$159.00 \times 0.30 = \$47.70$. You will save $47.70 if you buy the chair on sale.

Figure the amount of discount on each of the examples below. (DOK 2)

1. Hamburger is on sale for 20% off the normal $3.70 a pound price. How much can you save per pound buying the meat on sale?

2. Hardcover books are on sale for 15% off. Louise has chosen a hardcover book that normally sells for $20.00. How much will Louise save?

3. Alex chose a pair of jeans that sells for $19.50, but today all jeans are on sale for 10% off. How much will Alex save?

4. Dolly's Dress Shop is having a sale on dresses - buy one, get the second for half off (50%). Madison finds two dresses that normally sell for $50.00 each. How much will she save if she buys the two dresses during this special sale?

5. Henry finds a sale on tires. Henry buys four tires that normally sell for $40.00 each. If the tires are 25% off, how much will Henry pay for the 4 tires?

6. At a back to school sale for computers, Irwin buys one that normally costs $599.00. He receives the sale price at 10% off. How much does Irwin save on the computer. (Assume the sales tax is included in the original price of $599.00).

7. Mrs. Barkins is buying shoes for all six of her children. The total of the six pairs comes to $138.00. This week, the store is running a sale - 30% off all children's shoes. How much does she save buying the six pairs of shoes this week?

8. Mary Ellen has a coupon to get 25% off any sweater at the Yellow Bee Dress Shop. Mary Ellen chooses a sweater that costs $44.80, including tax. How much will Mary Ellen save with her coupon?

5.8   Finding the Discounted Sale Price (DOK 2)

To find the discounted sale price, you must go one step further than shown on the previous page. Read the example below to learn how to figure **discount** prices.

Example 11:   A $62.00 chair is on sale for 20% off. How much will it cost if I buy it now?

Step 1:   Change 20% to a decimal.

$20\% = 0.2$

Step 2:   Multiply the original price by the discount.

$$
\begin{array}{rr}
\textbf{ORIGINAL PRICE} & \$62.00 \\
\times \quad \textbf{\% DISCOUNT} & \times \quad 0.2 \\
\hline
\textbf{SAVINGS} & \$12.40
\end{array}
$$

Step 3:   Subtract the savings amount from the original price to find the sale price.

$$
\begin{array}{rr}
\textbf{ORIGINAL PRICE} & \$62.00 \\
- \quad \textbf{SAVINGS} & - \quad 12.40 \\
\hline
\textbf{SALE PRICE} & \$49.60
\end{array}
$$

Figure the sale price of the items below. The first one is done for you. (DOK 2)

ITEM	PRICE	%OFF	MULTIPLY	SUBTRACT	SALE PRICE
1. pen	$2.50	30%	$2.50 \times 0.3 = \$0.75$	$2.50 - 0.75 = 1.75$	$1.75
2. recliner	$420	50%			
3. juicer	$75	25%			
4. blanket	$16	30%			
5. earrings	$1.80	15%			
6. figurine	$12	20%			
7. boots	$84	10%			
8. calculator	$12	10%			
9. candle	$7.95	70%			
10. camera	$218	5%			
11. DVD player	$99.95	10%			
12. video game	$45	25%			

5.9   Markups (DOK 2)

A markup is the difference between the cost of a good and its selling price. It is often described as a percent increase. Businesses mark up their products in order to make a profit.

Example 12:   A pair of shoes has been marked up by 35%. It originally cost $15.00 to produce. What is the selling price of the shoes?

Step 1:   Change the percent to a decimal. $35\% = 0.35$

Step 2:   Find the amount the shoes have been marked up. $15 \times 0.35 = 5.25$

Step 3:   Add the markup to the cost. $5.25 + 15.00 = 20.25$

Answer: The selling price of the shoes is $20.25.

Example 13:   It costs a sporting goods company $20.81 to build a new skateboard. The company sells skateboards for $28.99 each. What is the percent markup on the skateboard to the nearest percent?

Step 1:   Find the difference between the cost and the price. $28.99 - 20.81 = 8.19$

Step 2:   Find the percent difference by dividing the difference by the cost. $\dfrac{8.19}{20.81} = 0.39$

Step 3:   Change the decimal to a percent. $0.39 = 39\%$

Answer: The percent markup on the skateboard is 39%.

Solve the following problems. (DOK 2)

1. It costs $15,700 to build a particular car. This car sells for $24,322. What is the percent markup on the car to the nearest percent?

2. It costs a florist $14.50 to grow a dozen roses. She sells a bouquet of a dozen roses for $24.99. How much does she mark up the flowers to the nearest percent?

3. It costs Jill $1.49 to make a glass of lemonade. She marks up her lemonade by 34%. For how much does she sell a glass of lemonade?

4. A gourmet chef uses $15.25 worth of ingredients to prepare a meal at an upscale restaurant. If he marks up his meal by 61%, what is the price of the meal?

5. A pair of jeans costs $24.50 to produce and is being sold for $34.50. What is the percent markup on the jeans, to the nearest percent?

6. It costs a farmer $1.74 to grow a pound of blueberries. The farmer marks up his blueberries by 50% to earn a profit. For how much does he sell a pound of blueberries?

5.10   Sales Tax (DOK 2)

Sales tax is a percentage added to the cost of goods and services that is paid to the government.

Example 14:   The total price of a sofa is $560.00 plus 6% **sales tax**. How much is the sales tax? What is the total cost?

   Step 1:   You will need to change 6% to a decimal.

   $6\% = 0.06$

   ☐ **Refer to the "About AR" on page xi!**

   Step 2:   Multiply the cost, $560, by the tax rate, 6%. $560 \times 0.06 = 33.6$
   The answer will be $33.60. (You need to add a 0 to the answer. When dealing with money, there must be two places after the decimal point.)

	COST		$560
$\times$	6% TAX	$\times$	0.06
	SALES TAX		$33.60

   Step 3:   Add the sales tax amount, $33.60, to the cost of the item sold, $560. This is the total cost.

	COST		$560.00
$+$	SALES TAX	$+$	33.60
	TOTAL COST		$593.60

Note: When the answer to the question involves money, you always need to round off the answer to the nearest hundredth (2 places after the decimal point). Sometimes you will need to add a zero.

Figure the total costs in the problems below. The first one is done for you. (DOK 2)

	ITEM	PRICE	% TAX	MULTIPLY	PRICE PLUS TAX	TOTAL
1.	jeans	$42	7%	$42 \times 0.07 = \$2.94$	$42 + 2.94 = 44.94$	$44.94
2.	truck	$17,495	6%			
3.	film	$5.89	8%			
4.	T-shirt	$12	5%			
5.	football	$36.40	4%			
6.	soda	$1.78	5%			
7.	4 tires	$105.80	10%			
8.	clock	$18	6%			
9.	burger	$2.34	5%			
10.	software	$89.95	8%			

5.11   Understanding Simple Interest (DOK 2)

I = PRT is a formula to figure out the **cost of borrowing money** or the **amount you earn** when you **put money in a savings account**. For example, when you want to buy a used truck or car, you go to the bank and borrow the $11,000 you need. The bank will charge you **interest** on the $11,000. If the simple interest rate is 4% for six years, you can figure the cost of the interest with this formula.

First, you need to understand these terms:

I = Interest = The amount charged by the bank or other lender
P = Principal = The amount you borrow
R = Rate = The interest the bank is charging you
T = Time = How many years you will take to pay off the loan

Example 15:   In the problem above: **I = PRT**. This means the **interest** equals the **principal**, times the **rate**, times the **time** in **years**.

$$I = \$11,000 \times 4\% \times 6 \text{ years}$$
$$I = \$11,000 \times 0.04 \times 6$$
$$I = \$2,640$$

Use the formula I = PRT to work the following problems. (DOK 2)

1. Craig borrowed $1,200 from his parents to buy a stereo. His parents charged him 5% simple interest for 3 years. How much interest did he pay his parents?

2. Raul invested $7,000 in a savings account that earned 11% simple interest. If he kept the money in the account for 8 years, how much interest did he earn?

3. Bridgette borrowed $15,000 to buy a car. The bank charged 13% simple interest for 5 years. How much interest did she pay the bank?

4. A tax accountant invested $35,000 in a money market account for 2 years. The account earned 12% simple interest. How much interest did the accountant make on his investment?

5. Linda Kay started a savings account for her nephew with $3,000. The account earned 5% simple interest. How much interest did the account accumulate in 4 months?

6. Renada bought a living room set on credit. The set sold for $1,700, and the store charged her 8% simple interest for three months. How much interest did she pay?

7. Duane took out a $2,450 loan at 7% simple interest for 2 years. How much interest did he pay for borrowing the $2,450?

5.12   Percent Error (DOK 2)

Percent error is the percent difference between an observed value and its actual value. The formula

for percent error is $\dfrac{|\text{observed value} - \text{true value}|}{\text{true value}}$.

Example 16: A student takes an object with an accepted mass of 200.00 grams and masses it on his own balance. He records the mass of the object as 196.5 g. What is his percent error?

Step 1: Find the difference between the two measurements. $196.5 - 200 = -3.5$

Step 2: Take the absolute value of the difference. $|3.5| = 3.5$

Step 3: Divide the result by the true value. $\dfrac{3.5}{200} = 0.0175$

Step 4: Change the decimal to a percent. $0.0175 = 1.75\%$

Answer: The percent error is 1.75%

Solve the following percent error problems. (DOK 2)

1. A carpenter needs a 1.75 ft long piece of wood to complete a new wood floor. He accidentally cuts a piece that is 1.83 ft long and realizes that it is too long when he tries to place it. What is the carpenter's percent error?

2. Jaden counts 50 cards in a deck that really has 52 cards in it. What is Jaden's percent error?

3. Mark made a mistake when measuring the volume of a big container. He found the volume to be 60 liters. However, the real value for the volume is 45 liters. What is the percent error?

4. Miss Sylvie counts 33 children on the bus before the class leaves for a school field trip. When she counts again, she realizes that there are actually only 30 students on the bus. What is her percent error?

5. Working in the laboratory, Tyler found the density of a piece of pure aluminum to be 2.87 g/cm^3. The accepted value for the density of aluminum is 2.699 g/cm^3. What is the student's percent error?

6. When estimating the product of 17 and 19, Joseph answers 300. The true product is 323. What is the percent error of Joseph's estimate?

5.13   Going Deeper into Percents (DOK 3)

While out shopping, you may see a sale sign that gives the percent off of a group of items. A quick way to find the sale price is to take the percent off and subtract it from 100%. This will give you the percent you will pay. Using **estimation**, you can calculate in your head how much you will pay for an item.

Example 17:   A shirt that normally costs $29.97 is on sale for 30% off.
Estimate the amount the shirt will cost on sale.

Step 1:   Subtract the percent off from 100%: $100\% - 30\% = 70\%$

Step 2:   Round the normal price of the shirt to the nearest **ten** dollars.
The nearest ten dollar is easier to multiply in your head.
$29.97 rounds to $30.

Step 3:   Convert 70% to a decimal and multiply by $30: $\$30 \times 0.7 = \$21.00$

Answer: On sale, the shirt costs about $21.00

Estimate the amount you will pay for each of the items below using the steps in the example above. Show your work for each step. (DOK 3)

1. video game costs $49.99, on sale for 40% off

2. coat costs $128.98, on sale for 50% off

3. book costs $19.95, on sale for 20% off

4. car costs $20,000, on sale for 30% off

5. flowers cost $29.99, on sale for 60% off

6. backpack costs $69.97, on sale for 40% off

A store is having a sale based on the number of items purchased using the chart below.

1st item - the highest priced item	20% off
2nd item - the second highest priced item	30% off
3rd item - the item that costs the least	50% item

Round the prices to the nearest ten dollars and then use the chart above to find the estimated sales prices. Find the total of the estimated prices of the three items each person purchased. Show your work for each step. (DOK 3)

7. Jean bought a shirt that costs $28.99, a pair of shoes that cost $39.99, and a bottle of perfume that cost $29.99.

8. Carlos bought a pair of jeans that cost $47.95, a belt that cost $10.99, and a pair of shoes that cost $58.95.

9. Isabelle bought a necklace that cost $20.99, a dress that cost $39.99, and a raincoat that cost $69.99.

Chapter 5 Review

Change the numbers according to the directions on each line. (DOK 1)

1. Change 41% to a decimal.

2. Change 0.315 to a percent.

3. Change 923% to a decimal.

4. Change 4.67 to a percent.

5. Change 92% to a fraction.

6. Change $\frac{2}{5}$ to a percent.

7. Change 478% to a fraction.

8. Change $2\frac{1}{2}$ to a percent.

Solve the word problems below. (DOK 2)

9. One hundred fifty students at Park City Middle School were polled and asked what is their favorite kind of movie. Twenty-two percent of the students polled said their favorite kind of movie was Science Fiction. How many students preferred Science Fiction movies?

10. The Wolverines basketball team won 12 games last season. This year, the team won 15 games. What percent increase did the Wolverines win this year over last year?

11. Chandra is a waitress at the French Cafe. She waited on a table of two gentlemen who purchased a total of $24.00 in food and ice tea, and left Chandra a tip of $3.60. What percent tip did Chandra receive?

12. Tim and his dad wanted to buy new T-shirts. There was a sale at the T-Shirt Factory Store. If you buy at least 3 T-shirts, you got 35% off your entire T-shirt purchase. Tim and his dad together purchased a total of 4 T-shirts for $8.00 each. How much did they save?

13. Laura has a coupon to get $1.00 off a $4.00 bottle of laundry detergent. What percent will Laura save?

14. Linda took out a simple interest loan for $8,000 at 11% interest for 3 years. How much interest did she have to pay back?

15. It costs a publishing company $10.15 to manufacture a test prep book. The book sells for $19.95. What is the percent markup on the book?

16. A bakery marks up its muffins by 40%. It costs $2.00 to make one banana nut muffin. What is the final price of the muffin?

17. There is a contest at school where students guess how many dimes are in a jar. The person whose guess is the closest to the actual value wins the jar. Jeffrey guessed 470 dimes. The actual value is 510. What is Jeffrey's percent error?

18. Susie measured the length of her pencil to be 6 inches long. Its actual length is 6.4 inches. What is the percent error?

Estimate the sale prices below to the nearest ten dollars. Then calculate the sales tax on the estimated sale price. Show your work for each step. (DOK 3)

19. The price of a sweatshirt is $28.95. It is on sale for 40% off. The sales tax is 6%.

20. The price of a refrigerator is $697.99. It is on sale for 20% off. The sales tax is 8%.

Chapter 5 Test

1 Change 94% to a decimal.

  A 0.94

  B 9.4

  C 94

  D 0.094

(DOK 1)

2 Change 6.22 to a percent.

  A 6.22%

  B 62.2%

  C 622%

  D 0.622%

(DOK 1)

3 Change 40% to a fraction.

  A $\dfrac{1}{20}$

  B $\dfrac{1}{40}$

  C $4\frac{1}{10}$

  D $\dfrac{2}{5}$

(DOK 1)

4 Change $6\frac{3}{4}$ to a percent.

  A 6.34%

  B 675%

  C 63.4%

  D 6.75%

(DOK 1)

5 Warren deposits $4.00 of his allowance each week in a savings account. He receives $16.00 each week if he completes his chores on the farm. What percent does Warren save?

  A 25%

  B 40%

  C 2.5%

  D 4.0%

(DOK 2)

6 Melea bought a shirt that was originally marked $28.00 for 20% off. How much did Melea pay for the shirt?

  A $20.00

  B $5.60

  C $22.40

  D $20.40

(DOK 2)

7 Liz is collecting all 24 dolls in a special collection. So far, she has received 18 of the dolls for her birthdays and special occasions. What percent of the collection does Liz have so far?

  A 75%

  B 18%

  C 24%

  D 50%

(DOK 2)

8 Colin's puppy, Squeeker, weighed 1.5 pounds at his first vet visit. When Squeeker was a year old, he weighed 7.5 pounds. What percent did Squeeker's weight increase between the two visits?

  A 500%

  B 300%

  C 400%

  D 600%

(DOK 2)

9 Breona had 4 batches of cookies she brought to a slumber party. When the girls were done eating, there was 1 batch of cookies left. What percent did the girls at the slumber party eat?

A 75%

B 50%

C 25%

D 7.5%

(DOK 2)

10 Change 0.01 to a fraction.

A $\dfrac{1}{10}$

B $\dfrac{10}{100}$

C $\dfrac{10}{1}$

D $\dfrac{1}{100}$

(DOK 1)

11 Jada earned a 7% commission on a $500 sale of a boat. How much was her commission?

A $35

B $3.50

C $350

D $0.35

(DOK 2)

12 A hat costs $17.00 and is on sale for 30% off. How much will be saved?

A $3.40

B $51.00

C $34.00

D $5.10

(DOK 2)

13 Mr. Allen had 8 hamsters in cages in his science class. The next year, Mr. Allen had 10 hamsters. What percent increase in hamsters did Mr. Allen have?

A 18%

B 20%

C 25%

D 2%

(DOK 2)

14 Will got a 25% tip on a meal costing $14.00. What was the amount of the tip?

A $3.50

B $3.00

C $2.50

D $2.80

(DOK 2)

15 Change $\dfrac{5}{8}$ to a percent.

A 6.25%

B 0.625%

C 62.5%

D 625%

(DOK 2)

16 What is 86% of 200?

A 172

B 86

C 286

D 114

(DOK 2)

17 Using the formula **I = PRT**, how much interest will you pay on a loan for $5,000 at 7% interest for 2 years?

A $350

B $700

C $140

D $190

(DOK 3)

18 Using the formula **I = PRT**, how much interest will you earn on $300 at 3% interest for 3 years?

A $2.70
B $90
C $9
D $27

(DOK 3)

19 It costs $2.10 for a gourmet coffee shop to make a small latte. The price of a small latte is $2.94. What is the percent markup on the latte?

A 25%
B 30%
C 35%
D 40%

(DOK 2)

20 A pair of sunglasses costs $14.50 to manufacture and is marked up by 32%. What is the selling price of the sunglasses?

A $19.14
B $19.43
C $20.30
D $20.59

(DOK 2)

21 During a chemistry experiment, Kevin calculated the molar mass of sodium to be 23.4 g. The actual molar mass of sodium is 22.99g. What is his percent error?

A 1.31%
B 1.55%
C 1.78%
D 2.02%

(DOK 2)

22 In geometry, Maddie measured the circumference of a circle to be 13 cm. The true value for the circumference is 12.566 cm. What is the percent error?

A 3.45%
B 4.36%
C 4.57%
D 4.71%

(DOK 2)

23 Using the formula **I = PRT**, how much interest will you earn on $300 at 3% interest for 3 years?

A $2.70
B $90
C $9
D $27

(DOK 3)

24 Change 16% to a decimal.

A 0.16
B 1.6
C 16.0
D 0.016

(DOK 2)

25 Estimate the sales price to the nearest ten dollars on a book that costs $39.98, is on sale for 40% off, and the sales tax is 5%.

A $24.00
B $25.20
C $24.20
D $16.80

(DOK 3)

26 Estimate the sales price to the nearest ten dollars on a bicycle that costs $129.98, is on sale for 30% off, and the sales tax is 7%.

A $41.73
B $92.57
C $97.37
D $98.37

(DOK 3)

Chapter 6
Rates, Ratios, and Proportions

This chapter covers the following CC Grade 7 standards:

	Content Standards
Ratios and Proportional Relationships	7.RP.1, 7.RP.3

6.1   Rate (DOK 2)

Example 1:   Laurie traveled 312 miles in 6 hours. What was her average rate of speed?

Divide the number of miles by the number of hours. $\dfrac{312 \text{ miles}}{6 \text{ hours}} = 52 \text{ miles/hour}$

Laurie's average rate of speed was 52 miles per hour (or 52 mph).

Find the average rate of speed in each problem below. (DOK 2)

1. A race car went 500 miles in 4 hours. What was its average rate of speed?

2. Carrie drove 124 miles in 2 hours. What was her average speed?

3. After 7 hours of driving, Chad had gone 364 miles. What was his average speed?

4. Anna drove 360 miles in 8 hours. What was her average speed?

5. After 3 hours of driving, Paul had gone 183 miles. What was his average speed?

6. Nicole ran 25 miles in 5 hours. What was her average speed?

7. A train traveled 492 miles in 6 hours. What was its average rate of speed?

8. A commercial jet traveled 1,572 miles in 3 hours. What was its average speed?

9. Jillian drove 195 miles in 3 hours. What was her average speed?

10. Greg drove 336 miles away from his home for 8 hours. What was his average speed?

11. Caleb drove 128 miles in two hours. What was his average speed in miles per hour?

12. After 9 hours of driving, Kate had traveled 405 miles. What speed did she average?

6.2   More Rates (DOK 2)

Rates are often discussed in terms of miles per hour, but a rate can be any measured quantity divided by another measurement such as feet per second, kilometers per minute, mass per unit volume, etc. A rate can be how fast something is done. For example, a bricklayer may lay 80 bricks per hour. Rates can also be used to find measurements such as density. For example, 35 grams of salt in 1 liter of water gives the mixture a density of 35 grams/liter.

Example 2:   Nathan entered his snail in a race. His snail went 18 feet in 6 minutes. How fast did his snail move?

In this problem, the units given are feet and minutes, so the rate will be feet per minute (or feet/minute).

You need to find out how far the snail went in one minute.

$$\text{Rate equals } \frac{\text{distance}}{\text{time}} \text{ so } \frac{18 \text{ feet}}{6 \text{ minutes}} = \frac{3 \text{ feet}}{1 \text{ minute}}$$

Nathan's snail went an average of 3 feet per minute or $3\dfrac{\text{ft}}{\text{min}}$

Find the average rate for each of the following problems. (DOK 2)

1. Tewanda read a 2,000 word news article in 8 minutes. How many words did she read per minute?

2. Chandler rides his bike to school every day. He travels 2,560 feet in 640 seconds. How many feet does he travel per second?

3. Mr. Molier is figuring out the semester averages for his history students. He can figure the average for 20 students in an hour. How long does it take him to figure the average for each student?

4. In 1908, John Hurlinger of Austria walked 1,400 kilometers from Vienna to Paris on his hands. The journey took 55 days. What was his average speed per day?

5. Spectators at the Super Circus were amazed to watch a cannon shoot a clown 212 feet into a net in 4 seconds. How many feet per second did the clown travel?

6. Marcus Page, star receiver for the Big Bulls, was awarded a 5 year contract for 105 million dollars. How much will his annual rate of pay be if he is paid the same amount each year?

7. Duke Delaney scored 28 points during the 4 quarters of the basketball playoffs. What was his average score per quarter?

8. The new McDonald's in Moscow serves 11,208 customers during a 24 hour period. What is the average number of customers served per hour?

9. Jimmy walks $\frac{1}{2}$ mile in $\frac{1}{6}$ hours. How fast does he walk in miles/hour?

10. A cheetah typically runs 15 miles in $\frac{1}{5}$ hours. How fast does a cheetah run in miles/hour?

11. A Japanese bullet train can move $\frac{1}{2}$ km in 6 seconds. How fast can the train move in km/hr?

12. A snail can move $\frac{1}{30}$ mile in $\frac{1}{3}$ hour. How fast does a snail move in miles/hour?

6.3   Ratio Problems (DOK 2)

In some word problems, you may be asked to express answers as a **ratio**. Ratios can look like fractions. Numbers must be written in the order they are requested. In the following example, 8 cups of sugar are mentioned before the 6 cups of strawberries. But in the question part of the example, you are asked for the ratio of STRAWBERRIES to SUGAR. The amount of strawberries IS THE FIRST WORD MENTIONED, so it must be the **top** number of the fraction. The amount of sugar, IS THE SECOND WORD MENTIONED, must be the **bottom** number of the fraction.

Example 3:   The recipe for jam requires 8 cups of sugar for every 6 cups of strawberries. What is the ratio of strawberries to sugar in this recipe?

First number requested $\dfrac{6}{8}$ $\dfrac{\text{cups strawberries}}{\text{cups sugar}}$
Second number requested

Answers may be reduced to lowest terms. $\dfrac{6}{8} = \dfrac{3}{4}$

This ratio could also be expressed as $3 : 4$.

Practice writing ratios for the following word problems and reduce to lowest terms. DO NOT CHANGE ANSWERS TO MIXED NUMBERS. Ratios should be left in fraction form. (DOK 2)

1. Out of the 301 seniors, 117 are boys. What is the ratio of boys to the total number of seniors?

2. A skyscraper that stands 530 feet tall casts a shadow that is 84 feet long. What is the ratio of the shadow to the height of the skyscraper?

3. Twenty boxes of paper weigh 460 pounds. What is the ratio of boxes to pounds?

4. The newborn weighs 7 pounds and is 21 inches long. What is the ratio of weight to length?

5. Jack paid $4.00 for 6 pounds of apples. What is the ratio of the price of apples to the pounds of apples?

6. It takes 11 cups of flour to make 3 loaves of bread. What is the ratio of cups of flour to loaves of bread?

6.4   Expressing Ratios as Percents and Percents as Ratios (DOK 2, 3)

Sometimes when you are doing ratio problems, you will be asked to find the percentage instead of the ratio. In the previous example, we found the ratio of strawberries to sugar. The ratio was $3 : 4$. Now we are going to find the percentage.

Example 4:   About what percent of the jam is strawberries?

Step 1:   The ratio is $3 : 4$, and that means for every 3 cups of strawberries there are 4 cups of sugar. Next, figure out how many cups there are total. To do this, add the two parts of the ratio: $3 + 4 = 7$. The ratio includes a <u>total</u> of 7 cups of ingredients.

Step 2:   The question asks to find the percent of the jam that is strawberries. Out of the 7 cups of ingredients, 3 of those cups are strawberries. Set up a fraction, and change that fraction into a percent.

$$\frac{3 \leftarrow \text{This is how many cups of strawberries are in \textbf{part} of the ratio.}}{7 \leftarrow \qquad \text{This is how many cups are in the \textbf{whole} ratio.}}$$

Step 3:   Change $\frac{3}{7}$ to a percent. To do this, change the fraction to a decimal by dividing.

$$\frac{3}{7} \approx 0.43 \; (3 \div 7 \approx 0.43)$$

Change the decimal answer, 0.43, to a percent by moving the decimal over 2 spaces to the right and adding a percent (%) sign to the end of the number. $0.43 = 43\%$

Answer: The jam is about 43% strawberries.

Example 5:   A restaurant decided to start serving milkshakes and took a survey on which flavor people preferred the most out of vanilla, chocolate, and strawberry. After 300 people were surveyed, the results were that 30% like vanilla, 45% like chocolate, and the rest of the people like strawberry. What is the ratio of the people who like strawberry to the people who like vanilla?

Step 1:   With the given percentages of the choices, find out what percent of the people like strawberry. To do this, add the two percent values together.

$$30\% + 45\% = 75\%$$

This means that 75% of the people surveyed like either vanilla or chocolate. The rest of the people must have preferred strawberry.

Subtract 75% from 100%.

$$100\% - 75\% = 25\%$$

25% of the people surveyed like strawberry.

Step 2:   Now find how many people each percent represents. To do this, change each percent to a decimal by moving the decimal two places to the left.

$$30\% = 0.30$$
$$45\% = 0.45$$
$$25\% = 0.25$$

Step 3:   Then multiply the total number of people surveyed, 300, by each decimal.

Vanilla: $300 \times 0.30 = 90$

Chocolate: $300 \times 0.45 = 135$

Strawberry: $300 \times 0.25 = 75$

Step 4:   Set up a ratio of the people who like strawberry to the people who like vanilla and simplify the ratio.

$$75 : 90 = 5 : 6$$

Answer: The ratio of the people who like strawberry to the people who like vanilla is $5 : 6$.

Read the following scenario carefully and answer the following questions. (DOK 2)

Fleta owns a candy store. On Monday, she sold 6 pounds of chocolate, 8 pounds of jelly beans, 5 pounds of sour snaps, and 6 pounds of yogurt-covered raisins.

1. What is the ratio of sour snaps sold to the total amount of candy sold?

2. What is the ratio of chocolate sold to jelly beans sold?

3. What is the ratio of yogurt-covered raisins sold to chocolate sold?

4. What is the ratio of chocolate and jelly beans sold to the total amount of candy sold?

5. What is the ratio of the candy sold that is <u>not</u> chocolate to the total amount of candy sold?

6. What is the ratio of sour snaps to chocolate?

Carefully read and answer the following questions. Explain your reasoning and show your work for each step. (DOK 3)

7. In a bag of marbles, the ratio of blue to red marbles is 2 : 3. Assuming that those are the only two colors in the bag, what is the percent of blue marbles in the bag?

8. What is the percent of red marbles in the bag in Question 1?

9. Students got to choose what game they wanted to play in gym class. The ratio of people wanting to play kickball to dodgeball was 7 : 5. What percent of the students wanted to play dodgeball? (Round to the nearest percent.)

10. A teacher took a survey on what the students' favorite color was between blue and green. The ratio of the two colors was 5 : 7. What percent of the students preferred blue? (Round to the nearest percent.)

11. The ratio of the number of students playing basketball to soccer at a local middle school is 5 : 2. What percent of the students at the middle school play soccer? (Round to the nearest percent.)

12. The ratio of brown socks to black socks in Camille's drawer is 7 : 3. Assuming that these are the only two color socks she has in her drawer, what is the percent of brown socks in Camille's drawer?

13. Marcus got a ratio of correctly answered questions to total questions of 9 : 12 on his past quiz. What percent of the quiz did Marcus answer incorrectly?

6.5 Solving Proportions (DOK 1)

Two **ratios (fractions)** that are **equal** to each other are called **proportions.** For example, $\frac{1}{4} = \frac{2}{8}$. Notice that the cross products are equal: $(4)(2) = (1)(8)$. **Read the following example to see how to find a number missing from a proportion.**

Example 6: $\qquad \frac{5}{15} = \frac{8}{x}$

Step 1: To find x, you first multiply the two numbers that are diagonal to each other.

$$\frac{5}{\{15\}} = \frac{\{8\}}{x}$$

$15 \times 8 = 120$

$5 \times x = 5x$

Therefore, $5x = 120$.

Step 2: Then solve for x by dividing the product (120) by the other number in the proportion (5).

$120 \div 5 = 24$

Therefore, $\frac{5}{15} = \frac{8}{24}$ **and** $x = 24$.

Practice finding the number missing from the following proportions. First, multiply the two numbers that are diagonal from each other. Then solve for x by dividing by the other number. (DOK 1)

1. $\frac{2}{4} = \frac{9}{x}$

2. $\frac{9}{3} = \frac{x}{7}$

3. $\frac{x}{12} = \frac{3}{6}$

4. $\frac{2}{x} = \frac{4}{12}$

5. $\frac{15}{x} = \frac{5}{3}$

6. $\frac{8}{x} = \frac{2}{5}$

7. $\frac{14}{6} = \frac{x}{3}$

8. $\frac{1}{x} = \frac{8}{64}$

9. $\frac{8}{2} = \frac{x}{3}$

10. $\frac{16}{2} = \frac{x}{4}$

11. $\frac{5}{6} = \frac{35}{x}$

12. $\frac{2}{x} = \frac{3}{18}$

13. $\frac{x}{4} = \frac{4}{16}$

14. $\frac{2}{5} = \frac{x}{40}$

15. $\frac{8}{4} = \frac{16}{x}$

16. $\frac{x}{2} = \frac{7}{14}$

17. $\frac{6}{12} = \frac{x}{8}$

18. $\frac{x}{40} = \frac{5}{20}$

19. $\frac{4}{8} = \frac{x}{4}$

20. $\frac{1}{4} = \frac{42}{x}$

6.6   Ratio and Proportion Word Problems (DOK 2, 3)

You can use ratios and proportions to solve problems.

Example 7:   A stick one meter long is held perpendicular to the ground and casts a shadow 0.4 meters long. At the same time, an electrical tower casts a shadow 112 meters long. Use ratio and proportion to find the height of the tower.

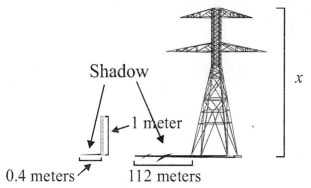

Step 1:   Set up a proportion using the numbers in the problem. Put the shadow lengths on one side of the equation and put the heights on the other side. The 1 meter height is paired with the 0.4 meter length, so let them both be top numbers. Let the unknown height be x.

$$\underset{\text{length}}{\overset{\text{shadow}}{\frac{0.4}{112}}} = \underset{\text{height}}{\overset{\text{object}}{\frac{1}{x}}}$$

Step 2:   Solve the proportion as you did on page 71.

$$112 \times 1 = 112 \qquad 112 \div 0.4 = 280$$

Answer: The tower height is 280 meters.

Use ratios and proportions to solve the following problems. (DOK 2, 3)

1. If 4 pounds of jelly beans cost $6.82, how much would 2 pounds cost?

2. Ashley drove from Memphis, Tennessee to Atlanta, Georgia. She drove for 5 hours and 45 minutes and drove 391 miles. How far did she travel in one hour?

3. Out of every 6 students surveyed, 1 listens to country music. At that rate, how many students in a school of 1,200 listen to country music?

4. Bailey, a Labrador retriever, has a litter of 10 puppies. Four are black. At that rate, how many puppies in a litter of 5 would be black?

5. According to the instructions on a bag of fertilizer, 5 pounds of fertilizer are needed for every 50 square feet of lawn. How many square feet will a 15-pound bag cover?

6. Faye wants to know how tall her school building is. On a sunny day, she measures the shadow of the building to be 8 feet. At the same time she measures the shadow cast by a 4 foot statue to be 2 feet. How tall is her school building?

7. For the first 3 home football games, the concession stand sold a total of 600 hotdogs. If that ratio stays constant, how many hotdogs will sell for all 8 home games?

8. A race car can travel 2 laps in 4 minutes. At this rate, how long will it take the race car to complete 250 laps?

9. If it takes 8 cups of flour to make 4 loaves of bread, how many loaves of bread can you make from 40 cups of flour?

10. Rudolph can mow a lawn that measures 1,000 square feet in 3 hours. At that rate, how long would it take him to mow a lawn 4,500 square feet?

11. Kathryn goes 12 mph when she rides her bike. How far does she go in 25 minutes?

12. Danny can paint $4\frac{1}{2}$ rooms in 3 hours. How many rooms can he paint in 5 hours?

13. Casey can do 4 math problems in $1\frac{3}{4}$ minutes. How many can he do in 7 minutes?

14. Sarah reads $1\frac{1}{4}$ pages in $1\frac{1}{2}$ minutes. How many pages does she read in 2 minutes?

15. Cameron runs $1\frac{1}{3}$ miles in 8 minutes. How many miles can he run in 20 minutes?

16. Jessie can swim 75 meters in $1\frac{3}{4}$ minutes. How far can he swim in 5 minutes?

6.7   Mixture Problems (DOK 3)

Ratios and proportions can often be used to figure out the concentration or mixture of certain problems.

Example 8:   A certain bleach mixture says that for every $\frac{1}{4}$ cup of bleach used there needs to be one gallon of water. How much bleach is needed to make a bleach mixture that calls for five gallons of water?

Step 1:   Set up a proportion.

📱 **Refer to the "About AR" on page xi!**

$$\frac{\frac{1}{4}}{x} = \frac{1}{5}$$

Step 2:   Cross multiply and solve for x.

$$\left(5 \times \frac{1}{4}\right) = (1 \times x) =$$

$$\frac{5}{4} = 1\tfrac{1}{4} = x$$

Answer: For a bleach mixture with five gallons of water, $1\tfrac{1}{4}$ cups of bleach is needed.

Solve the following mixture and concentration problems. Show your work. Simplify fractions when possible. (DOK 3)

1. For every 10,000 gallons of pool water, there needs to be 1 gallon of liquid chlorine. Steve has a 55,000 gallon pool. How much liquid chlorine does he need to add to his pool?

2. There needs to be 1 cup of sugar in every 2 quarts of fruit drink. How much sugar will someone need that wants to make 5 gallons of the fruit drink? (4 quarts = 1 gallon)

3. A bag of fertilizer covers 25 square feet of lawn. How many bags of fertilizer will Janice need if she has a $131\tfrac{1}{4}$ square foot lawn?

4. A humming bird feeder calls for $\frac{1}{4}$ cup of the nectar for every $\frac{2}{5}$ of a cup of water. How many cups of water is needed for a humming bird feeder that needs 1 cup of the nectar?

5. Making oatmeal needs $\frac{1}{2}$ cup of oatmeal per cup of water. At this rate, how many cups of water is needed for 18 cups of oatmeal?

6. A chemistry class is making salt water. For every 5 gallons of water they add 4 pounds of salt. How much salt will they need in order to make a 20 gallon salt water mixture with the same concentration?

Chapter 6 Review

Answer the following ratio and proportion questions. (DOK 2, 3)

1. Out of 250 coins, 50 are in mint condition. What is the ratio of mint condition coins to the total number of coins?

2. The ratio of boys to girls in the ninth grade is 6 : 5. If there are 185 girls in the class, how many boys are there?

3. Forty-six out of the total 310 seniors graduate with honors. What is the ratio of seniors graduating with honors to the total number of seniors?

4. Aunt Bess uses 3 cups of oatmeal to bake 6-dozen oatmeal cookies. How many cups of oatmeal would she need to bake 18-dozen cookies?

5. On a map, 3 centimeters represents 150 kilometers. If a line between two cities measures 4.5 centimeters, how many kilometers apart are they?

6. Shondra used four ounces of chocolate chips to make two dozen cookies. At that rate, how many ounces of chocolate chips would she need to make five dozen cookies?

7. When Rick measures the shadow of a yard stick, it is 7 inches. At the same time, the shadow of the tree he would like to chop down is 42 inches. How tall is the tree in yards?

8. The ratio of boys to girls in the gym class at a local middle school is 3 : 2. What percent of the gym class is girls?

9. There is a bag of marbles that has 40% blue marbles, 10% red marbles, 30% green marbles, and 20% black marbles. What is the ratio of red marbles to black marbles?

10. Johnny is looking at a map and measures the distance between Chattanooga and Nashville to be 3.5 inches. He wants to know the actual distance between the two cities. According to the scale on the map, 1 inch = 38 miles. How many miles is there between Chattanooga and Nashville?

11. Gorrest Fump runs 4 miles then walks 1 mile. How many miles does he run in every 10 miles?

Solve the following proportions. (DOK 2)

12. $\dfrac{4}{x} = \dfrac{1}{3}$

13. $\dfrac{3}{6} = \dfrac{x}{10}$

14. $\dfrac{x}{6} = \dfrac{4}{3}$

15. $\dfrac{2}{9} = \dfrac{6}{x}$

16. $\dfrac{9}{x} = \dfrac{3}{5}$

17. $\dfrac{15}{x} = \dfrac{1}{5}$

18. $\dfrac{7}{17} = \dfrac{x}{850}$

19. $\dfrac{x}{15} = \dfrac{5}{3}$

20. $\dfrac{3}{8} = \dfrac{x}{24}$

Chapter 6 Test

1 Gorden collects rocks. He has 35 sedimentary rocks, 25 igneous rocks, 15 metamorphic rocks, and 10 unknown rocks. Which ratio compares Gorden's sedimentary rocks to his unknown rocks, expressed in simplest form?

A $\frac{35}{10}$

B $35 : 10$

C $7 : 2$

D $2 : 7$

(DOK 2)

2 Applewood Elementary boasts a student to teacher ratio of $14 : 1$. What percent of the school makes up the number of teachers at Applewood Elementary? (Round percent to the nearest tenth.)

A 6.7%

B 59%

C 82.9%

D 4.5%

(DOK 2)

3 Find n: $\frac{3}{4} = \frac{75}{n}$.

A $n = 100$

B $n = 10$

C $n = 50$

D $n = 25$

(DOK 2)

4 The ratio of a rectangle's length to its width is $2 : 1$. If the length is 14 cm, what is its width?

A 14 cm

B 10 cm

C 7 cm

D 28 cm

(DOK 2)

5 A bag of marbles has 24 marbles. 37.5% are red marbles, 41.7% are green marbles, 12.5% are yellow marbles, and 8.3% are pink marbles. What is the ratio of the total number of marbles to yellow marbles? Express your answer in simplest terms.

A $\frac{24}{21}$

B $8 : 1$

C $8 : 7$

D $1 : 8$

(DOK 3)

6 Solve for x: $\frac{11}{9} = \frac{121}{x}$

A 11

B 12

C 0.81

D 99

(DOK 2)

7 An air tank that is 500 mL is 80% oxygen and 20% nitrogen. What is the amount of oxygen in milliliters in a 200 mL air tank that contains the same ratio?

A 160 mL

B 40 mL

C 400 mL

D 100 mL

(DOK 3)

8 Chris bought six pounds of ground beef for $10.80. How much would 15 pounds of ground beef cost at the same price per pound?

A $10.80

B $27.00

C $21.60

D $162.00

(DOK 2)

9 In a sand mixture there are 30 grams of dirt per 100 grams of mixture. Which proportion could be used to show how to find how much dirt is in a sand mixture of the same concentration that is 160 grams?

A $\dfrac{30}{100} = \dfrac{160}{x}$

B $\dfrac{160}{100} = \dfrac{30}{x}$

C $\dfrac{100}{30} = \dfrac{160}{x}$

D $\dfrac{x}{30} = \dfrac{100}{160}$

(DOK 2)

10 Solve for y: $\dfrac{y}{9} = \dfrac{63}{81}$

A 5

B 7

C 81

D 9

(DOK 2)

11 In a cookie jar there are 12 chocolate chip cookies, 3 peanut butter cookies, and 9 coconut cookies. In the neighbor's house there is a cookie jar that has the same proportion of cookies of the same kinds. If this cookie jar has 12 peanut butter cookies, how many chocolate chip cookies does the neighbor's cookie jar contain?

A 6

B 12

C 3

D 48

(DOK 3)

12 A box of greeting cards has 32 cards. 50% are birthday cards, 18.75% are get well cards, 25% are generic cards, and 6.25% are thank you cards. What is the ratio of the total number cards to get well cards? Express your answer in simplest terms.

A 16 : 3

B 8 : 1

C 8 : 3

D 16 : 1

(DOK 3)

Chapter 7
Proportional Relationships

This chapter covers the following CC Grade 7 standard:

	Content Standard
Ratios and Proportional Relationships	7.RP.2

7.1 Proportional Relationships (DOK 1)

One type of relationship between two sets of data is called a directly proportional relationship. When shown in a graph, this relationship can be drawn as a straight line. Direct relationships have a **positive slope** meaning that when the x values **increase** the y values **increase** at a constant rate, or can have a **negative slope** meaning that when the x values **increase** the y values **decrease** at a constant rate.

Direct relationships occur in a function when y varies directly, or in the same way, as x varies. The two values vary by a proportional factor, k. The variation is treated just like a proportion. The graphs shown below represent functions where x varies with y directly.

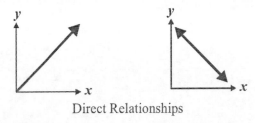

Direct Relationships

Example 1:   Direct relationships can also be demonstrated with function tables.

Table 1

x	y
0	3
1	4
2	7
3	12
4	19

Table 2

x	y
0	20
1	18
2	16
3	14
4	12

Notice in the first table, as x increases, y increases also. This means the first table represents a positive direct relationship between x and y. On the other hand, the second table shows a decrease in y when x increases. This means the second table represents a negative direct relationship between x and y.

7.2   Finding Relationships in Tables (DOK 2)

When looking at tables of data, there may be a relationship between the x (input) and y (output) values. Relationships can be linear (directly proportional), nonlinear (exponential, quadratic, etc.), or have no relationship.

Example 2:   What best describes the type of relationship shown between the input and the output values in the table below?

Input (x)	Output (y)
-1	-3
0	0
1	3
2	6
3	9

Step 1:   Look at the pairs of input and output in the table and see if there is a pattern.

Step 2:   Each output value is 3 times bigger than its paired input value. Also, every output value is increasing as well as increasing input values. This means that the values have a directly proportional relationship.

Answer: The relationship between the input and output values is linear and directly proportional.

What best describes the type of relationship shown between the two variables: directly proportional or not proportional? (DOK 2)

1.

x	y
1	8
3	18
5	28
9	48
10	53
20	103

3.

v	w
1	-2
5	-10
7	-14
9	-18
11	-22
13	-26

5.

f	g
1	5
3	-1
5	18
9	42
10	20
20	-52

2.

m	p
3	13
5	11
7	19
9	7
11	15
13	3

4.

x	y
-1	2
0	-5
1	8
2	-6
3	14
4	-19

6.

j	k
0	-3
2	0
4	3
6	6
8	9
10	12

7.3   Finding Relationships in Graphs (DOK 1)

Relationships between two variables can be found by looking at the graph of the two variables. The following example will show what a graph looks like with a linear relationship.

Example 3:   Look at the graph below and describe the relationship as either directly proportional with a positive slope or a negative slope.

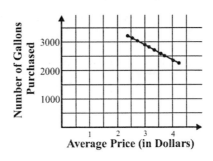

Step 1:   Recall on page 78 that a graph with x and y increasing together means it is directly proportional with a positive slope, and a graph with x increasing and y decreasing is directly proportional with a negative slope. This graph shows that as the average price (x) increases, the number of gallons purchased (y) decreases meaning that the relationship is directly proportional with a negative slope.

Answer: The relationship of the data in the graph is linear and directly proportional with a negative slope.

Look at the directly proportional relationships between the two variables in the graphs below, and decide what would best describe the relationship: Directly Proportional with a Positive Slope or Directly Proportional with a Negative Slope. (DOK 1)

1.
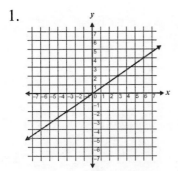

2.
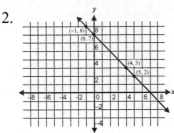

3.
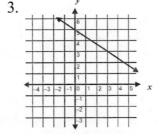

4.
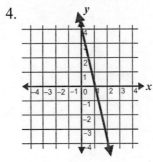

5.
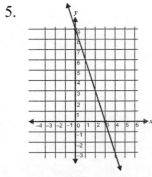

6.
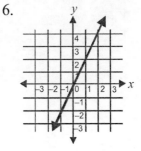

7.4   Finding Relationships in Equations (DOK 1)

Directly proportional relationships occur in a function when y varies directly with x. The two values vary by a proportional factor, k. Directly proportional relationships are expressed in the equation $y = kx$, where k is positive or negative.

Example 4:   What kind of relationship is represented by the equation $y = 5x + 2$?

Step 1:   Notice that the equation is in the same form as $y = kx$. This means that when the x values increase, the y values will increase in the same way or directly with x.

Answer: Since k is positive, 5, the relationship is linear and directly proportional.

Look at the following equations. State whether each directly proportional relationship has a positive slope or a negative slope. (DOK 1)

1. $y = 4x + 5$

2. $y = \frac{1}{2}x - 8$

3. $y = -9x + 15$

4. $y = 3.14159x + 4$

5. $y = 3 + \dfrac{x}{9}$

6. $y = 14x - 16$

7. $y = 6x + 15$

8. $y = 4 - 3x$

9. $y = 0.23 - x$

10. $y = 0.56 + \frac{1}{2}x$

11. $y = 1.23x - 56$

12. $y = \frac{4}{5}x + 9$

13. $y = \frac{11}{12}x - 1.56$

14. $y = -3x + 12$

7.5   Graphing Linear Data (DOK 3)

Many types of data are related by a constant ratio. As you learned on the previous page, this type of data is linear. The slope of the line described by linear data is the ratio between the data. Slope is the **constant rate of proportionality**. The **slope** is also known as the **unit rate**. Plotting linear data with a constant ratio can be helpful in finding additional values.

Example 5:   A department store prices socks by the pair. Each pair of socks costs $0.75. Plot pairs of socks versus price on a Cartesian plane.

Step 1:   Since the price of the socks is constant, you know that one pair of socks costs $0.75, 2 pairs of socks cost $1.50, 3 pairs of socks cost $2.25, and so on. Make a list of a few points.

Pair(s) x	Price y
1	0.75
2	1.50
3	2.25

Refer to the "About AR" on page xi!

Step 2:   Plot these points on a Cartesian plane and draw a straight line through the points.

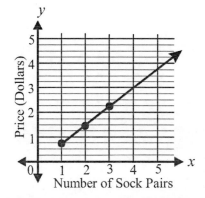

Example 6:   What is the unit rate of the data in the example above? What does the unit rate describe?

Answer: You can determine the unit rate either by the graph or by the data points. For this data, the slope is 0.75. Remember, unit rate is rise/run. For every $0.75 going up the y-axis, you go across one pair of socks on the x-axis. The unit rate describes the price per pair of socks.

Example 7:   Use the graph created in the example above to answer the following questions. How much would 5 pairs of socks cost? How many pairs of socks could you purchase for $3.00? Extending the line gives useful information about the price of additional pairs of socks.

Answer: The line that represents 5 pairs of socks on the x-axis intersects the data line at $3.75 on the y-axis. Therefore, 5 pairs of socks would cost $3.75. The line representing the value of $3.00 on the y-axis intersects the data line at 4 on the x-axis. Therefore, $3.00 will buy exactly 4 pairs of socks.

Example 8:   A movie theater sells movie tickets for $5.00 each. Plot the number of movie tickets versus price on a Cartesian plane. What is the unit rate of the data? What does the unit rate describe?

Step 1:   Since the price of movie tickets is constant, you know that one ticket costs $5.00, two tickets cost $10.00, three tickets cost $15.00, and so on. Make a list of a few points.

Tickets, x	Price, y
0	0
1	5
2	10
3	15

Step 2:   Plot these points on a Cartesian plane and draw a straight line through the points.

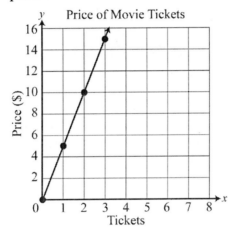

Notice that the line starts at the origin, $(0, 0)$. This is true because when x is 0, no tickets have been bought.

Step 3:   Determine the unit rate of the data.

By looking at the graph, you can see that the slope is 5. Since the unit rate is $\frac{\text{rise}}{\text{run}}$, the unit rate is 5.

The unit rate describes the price per ticket.

Notice that when x is 1, y is 5. In other words, when one movie ticket is bought, the price is $5.00. The unit rate is also $5.00, so when x is 1, y is equal to the unit rate. This is always true for data whose data line passes through the origin. Therefore, whenever $(0, 0)$ is a data point, the unit rate, k, is the y-coordinate for $x = 1$.

x	y
0	0
1	k

$$(0, 0) \rightarrow (1, k)$$

Use the information given to make a line graph for each set of data, and answer the questions related to each graph. (DOK 3)

1. The diameter of a circle versus the circumference of a circle is a constant ratio. Use the data given below to graph a line to fit the data. Extend the line, and use the graph to answer the next question.

Circle

Diameter	Circumference
2	6.28
3	9.42

2. Using the graph of the data in question 1, estimate the circumference of a circle that has a diameter of 4 inches.

3. If the circumference of a circle is 15.70 inches, about how long is the diameter?

4. What is the unit rate of the line you graphed in question 1?

5. What does the unit rate of the line in question 4 describe?

6. The length of a side on a square and the perimeter of a square are constant ratios to each other. Use the data below to graph this relationship.

Square

Length of side	Perimeter
4	16
5	20

7. Using the graph from question 6, what is the perimeter of a square with a side that measures 3 inches?

8. What does the unit rate of the line graphed in question 6 represent?

9. Conversions are often constant ratios. For example, converting from pounds to ounces follows a constant ratio. Use the data below to graph a line that can be used to convert pounds to ounces.

Measurement Conversion

Pounds	Ounces
3	48
4	64

10. Use the graph from question 9 to convert 32 ounces to pounds.

11. What does the unit rate of the line graphed for question 9 represent?

12. Graph the data below, and create a line that shows the conversion from weeks to days. What does the unit rate of the line represent?

Time

Weeks	Days
1	7
3	21

13. About how many days are in $3\frac{1}{2}$ weeks?

14. Graph a data line that converts feet to inches.

15. Using the graph in question 14, how many inches are in 3 feet?

16. What does the unit rate of the line of converting feet to inches represent?

17. A data line on a scatterplot intersects the following points; $(0,0)$, $(1,2)$, $(2,4)$. What is the unit rate of this data?

18. Using the data below, determine the unit rate and what it represents.

Feet	Seconds
0	0
1	3

7.6   Applying Proportional Relationships (DOK 3)

Proportional relationships can be applied to solve real life problems. Remember that a proportional relationship is a relationship between quantities. Some examples of these quantities are distance between places, length, weight, time, speed, price, height, and population.

Example 9:   A bee travels 50 meters in 2 seconds. At this speed, how long would it take for the bee to travel 375 meters?

Step 1:   Make a table displaying the time that has elapsed in seconds for the distance traveled in meters.

Time (seconds), x	Distance (meters), y
2	50
4	100
6	150
8	200

Step 2:   Determine the unit rate.

Use the formula $y = kx$ and a set of points from the table to solve for k.

Set of points: $x = 4$, $y = 100$

$$y = kx$$
$$100 = 4k \qquad \text{Substitute } x \text{ and } y \text{ into the formula.}$$
$$\frac{100}{4} = \frac{4k}{4} \qquad \text{Divide both sides of the equation by 4.}$$
$$25 = k \qquad \text{The unit rate is 25.}$$

Step 3:   Use the unit rate to find the time it would take to travel 375 meters.

$k = 25, y = 375$

$$y = kx$$
$$375 = 25x \qquad \text{Substitute } k \text{ and } y \text{ into the formula.}$$
$$\frac{375}{25} = \frac{25x}{25} \qquad \text{Divide both sides of the equation by 25.}$$
$$15 = x \qquad \text{The time elapsed is 15 seconds.}$$

Step 4:   Interpret the unit rate and the results.

The unit rate, $k = 25$, means that for every second that goes by, the bee travels 25 feet.

This means it would take the bee 15 seconds to travel 375 meters.

Answer: 15 seconds

Example 10:   A candy store sells 3 chocolate bars for $2.25. The data for the price of chocolate bars is represented in the graph below. Use this information to determine the number of chocolate bars Virginia could buy if she has $10.50.

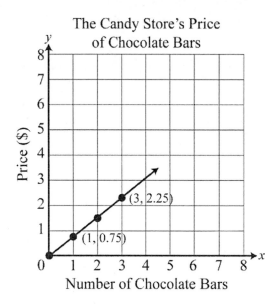

The Candy Store's Price of Chocolate Bars

Number of Chocolate Bars

Step 1:   Determine the unit rate.

Looking at the graph, we can see that the data line passes through the point $(0, 0)$. In the previous section, we learned a faster way to determine the unit rate when the data line passes through the origin. Remember if the point $(0, 0)$ is a point on the data line, the point $(1, k)$ is the unit rate.

The line on the graph shows that when $x = 1$, $y = 0.75$.

Therefore, the unit rate is $0.75.

Step 2:   Use the unit rate to determine the number of chocolate bars Virginia could buy with $10.50.

We know that $k = 0.75$ and $y = 10.50$.

$$
\begin{aligned}
y &= kx \\
10.50 &= 0.75x && \text{Substitute } y \text{ and } k \text{ into the formula.} \\
\frac{10.50}{0.75} &= \frac{0.75x}{0.75} && \text{Divide both sides of the equation by } 0.75. \\
14 &= x && \text{The number of chocolate bars is } 14.
\end{aligned}
$$

Answer: Virginia could buy 14 chocolate bars with $10.50.

Use the unit rate and proportional relationships to answer the following questions. (DOK 3)

1. Dr. Haley bought 15 boxes of cotton balls. He has a total of 525 cotton balls. How many cotton balls are there in each box?

2. Michael is doing his science homework and needs to know how many inches there are in 6.5 feet. Use the graph below, which represents the number of inches per foot, to determine the number of inches in 6.5 feet.

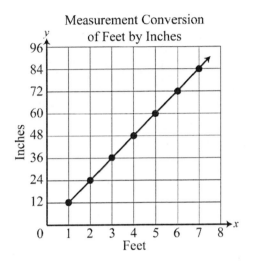

3. The Bryan family is traveling to their beach house 300 miles away. The longer they drive, the shorter the distance left to travel. After 30 minutes, there are 270 miles left to their destination. How many minutes have they driven if there are 180 miles left to their destination?

4. Jake is riding his bicycle at a constant rate. After 2 seconds he has gone 10 feet. How many feet did jake ride in 1 second? How many feet will he ride in 10 seconds?

5. At the local grocery store, apples are sold by the pound. The following table shows the total price for each pound of apples sold. What is the price for 5 pounds of apples?

Price of Apples	
Number of Pounds of Apples, x	Total Price, y
0	0
1	$3.00
2	$6.00
3	$9.00

7.7   Comparing Proportional Relationships (DOK 2)

As you have seen throughout this chapter, proportions can be represented in several ways, including tables, graphs, and equations. Study the following examples to determine if two proportions are equivalent.

Example 11:   Determine whether the following proportions are equivalent.

A.

x	y
2	6
3	9
4	12

B. $y = 3x$

Step 1:   Find the unit rate of the table in A.

As x increases by 1, y increases by 3. We can see this by dividing each y-coordinate by its corresponding x-coordinate.

$$\frac{9-6}{3-2} = \frac{12-9}{4-3} = \frac{3}{1} = 3$$

So, the unit rate of A is 3.

Step 2:   Find the unit rate of the equation in B. Remember that slope is the same as the unit rate. In the equation $y = 3x$, the slope is 3.

So, the unit rate of B is 3.

Answer: Since the unit rate of both A and B is 3, the proportions are equivalent.

Example 12:   Determine whether the proportion shown in the graph is equivalent to the proportion shown in the table.

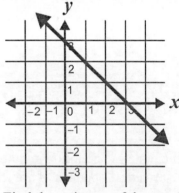

x	y
0	2
1	1
2	0
3	−1

Step 1:   Find the unit rate of the graph.

The slope of the line is −1, so the unit rate of the graph is −1.

Step 2:   Find the unit rate of the table.

$$\frac{1-2}{1-0} = \frac{0-1}{2-1} = \frac{-1-0}{3-2} = \frac{-1}{1} = -1$$

So, the unit rate is −1.

Answer: Since the unit rates of the table and graph are −1, the proportions are equivalent.

In numbers 1–8, match the data with the equivalent proportion (A–H). (DOK 2)

1. $y = \frac{1}{2}x + 2$

2.

x	y
0	0
1	6

3.

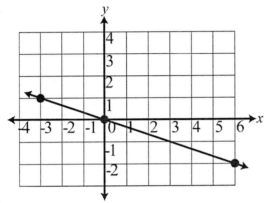

4. $y = -4x$

5.

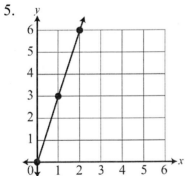

6.

x	y
1	6
2	4
3	2

7. $y = -5x + 3$

8.

x	y
−1	2
0	1
1	0
2	−1

(A)

x	y
1	−4
2	−8
3	−12

(B) $y = -\frac{1}{3}x$

(C)

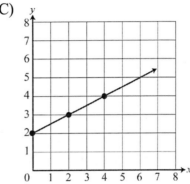

(D)

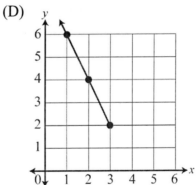

(E) $y = 6x$

(F)

x	y
0	0
1	3
2	6

(G) $y = -x + 1$

(H)

x	y
0	3
1	−2
2	−7
3	−12

Chapter 7 Review

Use the graph below to answer the next three questions. (DOK 2)

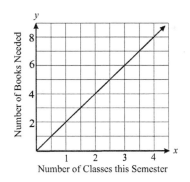

1. What is the best description of the type of relationship between the Number of Classes this Semester and Number of Books Needed?

2. What does the unit rate of the line in the graph represent?

3. How many books would be expected to be needed when taking 6 classes in a semester?

Answer the proportional relationship questions. (DOK 1, 2)

4. What type of relationship is involved when the y variable increases as the x variable increases?

5. What type of relationship is involved when the y variable decreases as the x variable increases?

6. What type of relationship **best** describes the equation $y = 4x - 2$?

7. What type of relationship best describes the relationship between x and y?

x	y
-2	4
-1	2
0	0
1	-2
2	-4

8. Which of the following data points represents the unit rate at the y-coordinate?

$(0, 0), (1, 2), (2, 4), (3, 6)$

9. Use the following graph to determine the unit rate.

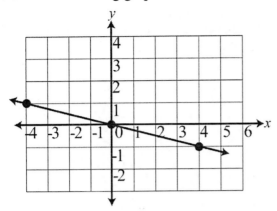

10. The shoe store sells shoes by the pair for $8.00 a pair. How much would it cost to buy 4 pairs of shoes?

11. Farmer John sells his tomatoes at his vegetable stand. His neighbor just bought 5 tomatoes for $2.50. What is the unit rate for the price of tomatoes?

12. Write an equation with proportions that are equivalent to the graph below. Then give two scenarios where this proportion may apply. **(DOK 3)**

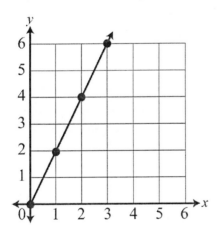

13. Write an equation with proportions that are equivalent to the table below. Give two scenarios where these proportions may apply. **(DOK 3)**

x	y
-2	6
-1	3
0	0
1	-3
2	-6
3	-9

Chapter 7 Test

1 What type of relationship best describes the relation in the graph below?

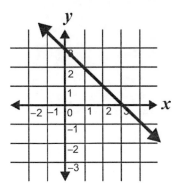

   A directly proportional with a negative slope

   B directly proportional with a positive slope

   C nonlinear

   D inversely proportional

(DOK 1)

2 What type of relationship best describes the relation in the table below?

x	y
0	3
2	1
3	0
4	−1

   A not proportional

   B directly proportional with a negative slope

   C nonlinear

   D directly proportional with a positive slope

(DOK 1)

3 What does the slope of the line in the graph below represent?

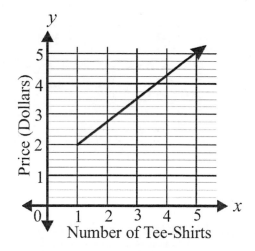

   A the price of a single tee-shirt

   B the price per tee-shirt

   C the number of tee-shirts

   D the price of all the tee-shirts

(DOK 2)

4 If it takes two men an hour to unload a truck full of merchandise, how long would it take six men to do the same amount of work?

   A 1 hour

   B 45 minutes

   C 20 minutes

   D 30 minutes

(DOK 2)

5 When $x = 4$, $y = 16$. What would be the value of y when $x = 9$ if this is a directly proportional relationship?

   A 36

   B 72

   C 3.6

   D 4

(DOK 2)

6 A book at a publishing company can typically get done in 2 months by a single person. How long would it take the publishing company to finish a single book if they split the work up among 4 people?

A 1 month

B 0.5 months

C 2 months

D 8 months

(DOK 2)

7 If the slope of a line in a graph is constant and the x and y values are increasing together at a constant rate, what best describes the type of relationship that is occurring?

A directly proportional with a negative slope

B directly proportional with a positive slope

C nonlinear

D inversely proportional

(DOK 1)

8 What is the unit rate of the data if two points are $(0,0)$ and $(1,4)$?

A 2

B 8

C 4

D 1

(DOK 2)

9 Jessi is buying baby clothes that come in a package of 3 dresses. Each package of 3 dresses costs $15.00. How much will it cost Jessi to buy 9 dresses?

A $40.00

B $27.00

C $30.00

D $45.00

(DOK 2)

10 Which of the following has the correct reasoning and is an equivalent proportion to the equation $y = -\frac{1}{2}x$?

A

x	y
0	0
1	$-\frac{1}{2}$

For every value of x, y is negative and half the size of x.

B

x	y
0	0
1	$-\frac{1}{2}$

For every value of x, x is negative and half the size of y.

C

x	y
1	$\frac{1}{2}$
4	2

For every value of x, y is negative and half the size of x.

D

x	y
0	0
1	$\frac{1}{2}$

For every value of x, x is negative and half the size of y.

(DOK 3)

11 Which of the following is an equivalent proportion to the graph below?

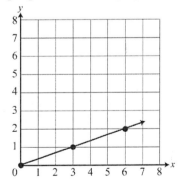

A $y = \frac{1}{2}x$

B $y = \frac{1}{3}x$

C $y = 3x$

D $y = -\frac{1}{3}x$

(DOK 2)

Chapter 8
Introduction to Algebra

This chapter covers the following CC Grade 7 standards:

	Content Standards
Expressions and Equations	7.EE.2, 7.EE.4

8.1   Algebra Vocabulary (DOK 1)

Vocabulary Word	Example	Definition
variable	$4x$ (x is the variable)	a letter that can be replaced by a number
coefficient	$4x$ (4 is the coefficient)	a number multiplied by a variable or variables
term	$5x^2 + x - 2$ ($5x^2$, x, and -2 are terms)	numbers or variables separated by $+$ or $-$ signs
constant	$5x + 2y + 4$ (4 is a constant)	a term that does not have a variable
degree	$4x^2 + 3x - 2$ (the degree is 2)	the largest power of a variable in an expression
leading coefficient	$4x^2 + 3x - 2$ (4 is the leading coefficient)	the number multiplied by the term with the highest power
sentence	$2x = 7$ or $5 \leq x$	two algebraic expressions connected by $=, \neq, <, >, \leq, \geq$, or $\approx$
equation	$4x = 8$	a sentence with an equal sign
inequality	$7x < 30$ or $x \neq 6$	a sentence with one of the following signs: $\neq, <, >, \leq, \geq$, or $\approx$
base	6^3 (6 is the base)	the number used as a factor
exponent	6^3 (3 is the exponent)	the number of times the base is multiplied by itself

8.2   Substituting Numbers for Variables (DOK 1, 2)

These problems may look difficult at first glance, but they are very easy. Simply replace the variable with the number the variable is equal to, and solve the problems.

Example 1:   In the following problems, substitute 3 for b.

Problem	Calculation	Solution
1. $b + 2$	Simply replace the b with 3. $3 + 2$	5
2. $13 - b$	$13 - 3$	10
3. $4b$	This means multiply. 4×3	12
4. $\dfrac{27}{b}$	This means divide. $27 \div 3$	9
5. b^3	$3 \times 3 \times 3$	27
6. $4b + 8$	$(4 \times 3) + 8$	20

Note: Be sure to do all multiplying and dividing before adding and subtracting.

Example 2:   In the following problems, let $a = 3$, $b = -1$, and $c = 4$.

Problem	Calculation	Solution
1. $3ac - b$	$3 \times 3 \times 4 - (-1)$	37
2. $cb^2 + 2$	$4 \times (-1)^2 + 2 = 4 \times 1 + 2$	6
3. $\dfrac{ac}{3}$	$(3 \times 4) \div 3 = 12 \div 3$	4

In the following problems let $r = 9$. Solve the problems. (DOK 1)

1. $r - 3 =$

2. $11 + r =$

3. $\dfrac{63}{r} =$

4. $r^2 + 1 =$

5. $4r - 6 =$

6. $r^2 - 80 =$

7. $\dfrac{r^2}{9} =$

8. $3r + 4 =$

9. $6r \div 2 =$

In the following problems let $r = -4$, $s = 10$, $t = 2$, and $u = -1$. Solve the problems. (DOK 2)

10. $-\frac{1}{4}r + \frac{1}{8}t =$

11. $\dfrac{tu}{9} =$

12. $s - 7 =$

13. $2ut + r =$

14. $\dfrac{3t}{2} =$

15. $u^2 + s + 1 =$

16. $3(4 + t) =$

17. $s - 9 + t =$

18. $\frac{3}{2}r^2 + 4 =$

19. $rtu =$

20. $3s + \frac{1}{5}sr =$

21. $rs + tu =$

8.3   Understanding Algebra Word Problems (DOK 1, 2)

The biggest challenge to solving word problems is figuring out whether to add, subtract, multiply, or divide. Below is a list of key words and their meanings. This list does not include every situation you might see, but it includes the most common examples.

Words Indicating Addition	**Example**	**Add**
and	3 **and** 9	$3 + 9$
increased	The original price of $14 **increased** by $2.	$14 + 2$
more	7 coins and 3 **more**	$7 + 3$
more than	Josh has 15 points. Will has 3 **more than** Josh.	$15 + 3$
plus	2 baseballs **plus** 1 baseballs	$2 + 1$
sum	the **sum** of 4 and 2	$4 + 2$
total	the **total** of 9, 5, and 11	$9 + 5 + 11$

Words Indicating Subtraction	**Example**	**Subtract**
decreased	$19 **decreased** by $7	$19 - 7$
difference	the **difference** between 24 and 10	$24 - 10$
less	12 days **less** 5	$12 - 5$
less than	Jose completed 11 laps **less than** Mike's 15.	$*15 - 11$
left	Ray sold 22 out of 40 tickets. How many did he have **left**?	$*40 - 22$
lower than	This month's rainfall is 3 inches **lower than** last month's rainfall of 9 inches.	$*9 - 3$
minus	8 **minus** 7	$8 - 7$

* In subtraction word problems, you cannot always subtract the numbers in the order that they appear in the problem. Sometimes the first number should be subtracted from the last. You must read each problem carefully.

Words Indicating Multiplication	**Example**	**Multiply**
triple	Her $150 profit **tripled** in in a month.	150×3
half	**Half** of the $800 collected went to charity.	$\frac{1}{2} \times 800$
product	the **product** of 5 and 11	5×11
times	Li scored 5 **times** as many points as Ted who only scored 3.	5×3
double	The bacteria **doubled** its original colony of 5,000 in just one day.	$2 \times 5,000$
twice	Ron has 8 CDs. Tom has **twice** as many.	2×8

Words Indicating Division	**Example**	**Divide**
divide into, by, or among	The group of 20 **divided into** 5 teams	$20 \div 5$ or $\frac{20}{5}$
quotient	the **quotient** of 36 and 4	$36 \div 4$ or $\frac{36}{4}$

Match the phrase with the correct algebraic expression below. The answers will be used more than once. (DOK 1)

A. $x + 4$ B. $4x$ C. $4 - x$ D. $x - 4$ E. $\dfrac{x}{4}$

1. 4 more than x

2. x divided into 4

3. 4 less than x

4. four times x

5. the quotient of x and 4

6. x increased by 4

7. 4 less x

8. the product of 4 and x

9. x decreased by 4

10. x times 4

11. 4 minus x

12. the total of 4 and x

Now practice writing parts of algebraic expressions from the following word problems. (DOK 2)

Example 3:   the product of 3 and a number, x Answer: $3x$

13. the sum of 3 and y

14. x minus 2

15. the quotient of r divided by 7

16. 5 more than p

17. 2 less than y

18. triple n

19. the total of h and 14

20. 7 less r

21. double y

22. 2 increased by c

23. 8 less than z

24. half of r

25. 4 times t

26. z minus 5

27. 8 plus m

28. 3 divided by s

29. the product of 4 and n

30. z decreased by 10

31. four times as much as x

32. q less than 12

If a word problem contains the word "sum" or "difference," put the numbers that "sum" or "difference" refer to in parentheses to be added or subtracted first. Do not separate them. Look at the examples below.

Examples:

Refer to the "About AR" on page xi!

	RIGHT	**WRONG**
sum of 2 and 4, times 5	$5(2+4) = 30$	$2 + 4 \times 5 = 22$
the sum of 4 and 6, divided by 2	$\dfrac{(4+6)}{2} = 5$	$4 + \dfrac{6}{2} = 7$
4 times the difference between 10 and 5	$4(10-5) = 20$	$4 \times 10 - 5 = 35$
20 divided by the difference between 4 and 2	$\dfrac{20}{(4-2)} = 10$	$20 \div 4 - 2 = 3$
the sum of x and 4, multiplied by 2	$2(x+4) = 2x + 8$	$x + 4 \times 2 = x + 8$

Change the following phrases into algebraic expressions. (DOK 2)

1. 4 times the sum of x and 2

2. the difference between 8 and 4, divided by 2

3. 60 divided by the sum of 5 and 2

4. twice the sum of 15 and x

5. the difference between x and 7, divided by 3

6. 6 times the difference between x and 3

7. 10 multiplied by the sum of 4 and 5

8. the difference between x and 3, divided by 5

9. x divided by the sum of 7 and 2

10. x minus 3, times 7

11. 70 multiplied by the sum of x and 4

12. twice the difference between 4 and x

13. 8 times the sum of 2 and 9

14. 3 times the difference between 8 and 1

15. 14 divided by the sum of 3 and 11

16. four minus x, multiplied by 15

Look at the examples below for more phrases that may be used in algebra word problems.

Examples

one-half of the sum of x and 4	$\frac{1}{2}(x+4)$ or $\dfrac{x+4}{2}$
six more than four times a number, x	$6+4x$
100 decreased by the product of a number, x, and 5	$100-5x$
ten less than the product of 3 and x	$3x-10$

Change the following phrases into algebraic expressions. (DOK 2)

1. one-third of the sum of x and 2

2. three more than the product of a number, x, and 5

3. ten less than the sum of t and 8

4. the product of 2 and n, minus 8

5. 9 less than the sum of 4 and x

6. the difference of the numbers 16 and 12, times a number, n

7. one-eighth of t

8. the product of 4 and x, minus 9

9. x times the difference between 2 and x

10. five plus the quotient of x and 9

11. the sum of 8 and k, divided by 3

12. one less than the product of 7 and x

13. 4 increased by one-half of a number, n

14. 25 more than twice x

15. seven subtracted from four times m

16. 9 times x, subtracted from 13

8.4   Setting Up Algebra Word Problems (DOK 2)

To complete an algebra problem, an equal sign must be added. The words **"is"** or **"are"** as well as **"equal(s)"** signal that you should add an equal sign.

Example 3:   <u>Double Jake's age</u>, x, <u>minus 4 is 22</u>.

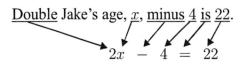

$$2x - 4 = 22$$

Translate the following word problems into algebra problems. <u>DO NOT</u> find the solutions to the problems <u>yet</u>. (DOK 2)

1. Triple the original number, n, is 3,500.

2. The product of a number, y, and 3 is equal to 15.

3. Four times the difference of a number, x, and 10 is 35.

4. The total, t, divided into 7 groups is 20.

5. The number of parts in inventory, p, minus 80 parts sold today is 270.

6. One-half an amount, x, added to $45 is $342

7. One hundred seeds divided by 4 rows equals n number of seeds per row.

8. A number, y, less than 20 is 47.

9. His base pay of $500 increased by his commission, x, is $640.

10. Seventeen more than half a number, h, is 100.

11. This month's sales of $2,900 are double January's sales, x.

12. The quotient of a number, w, and 8 is 24.

13. Six less a number, d, is 32.

14. Four times the sum of a number, y, and 7 is 84.

15. We started with x number of students. When 2 moved away, we had 28 left.

16. A number, b, divided by 29 is 3.

8.5   Changing Algebra Word Problems to Algebraic Equations (DOK 2)

Example 4:   There are 3 people who have a total weight of 595 pounds. Sally weighs 20 pounds less than Jessie. Rafael weighs 15 pounds more than Jessie. How much does Jessie weigh?

Step 1:   Notice everyone's weight is given in terms of Jessie. Sally weighs 20 pounds less than Jessie. Rafael weighs 15 pounds more than Jessie. First, we write everyone's weight in terms of Jessie, j.

$$\text{Jessie} = j$$
$$\text{Sally} = j - 20$$
$$\text{Rafael} = j + 15$$

Step 2:   We know that all three together weigh 595 pounds. We write the sum of everyone's weight equal to 595.

$$j + j - 20 + j + 15 = 595$$

We will learn to solve these problems in chapter 9.

Change the following word problems to algebraic equations. (DOK 2)

1. Fluffy, Spot, and Shampy have a combined age in dog years of 82. Spot is 14 years younger than Fluffy. Shampy is 6 years older than Fluffy. What is Fluffy's age, f, in dog years?

2. Jerry Marcosi puts 8% of the amount he makes per week into a retirement account, r. He is paid $12.00 per hour and works 40 hours per week for a certain number of weeks, w. Write an equation to help him find out how much he puts into his retirement account.

3. A furniture store advertises a 35% off liquidation sale on all items. What would the sale price (p) be on a $2,742 dining room set?

4. Kyle Thornton buys an item which normally sells for a certain price, x. Today the item is selling for 25% off the regular price. A sales tax of 8% is added to the equation to find the final price, f.

5. Tamika Francois runs a floral shop. On Tuesday, Tamika sold a total of $800 worth of flowers. The flowers cost her $75, and she paid an employee to work 8 hours for a given wage, w. Write an equation to help Tamika find her profit, p, on Tuesday. (This profit will be adjusted as Tamika will still need to pay herself, rent, utilities, etc. from the sales for the day.)

6. Sharice is a waitress at a local restaurant. She makes an hourly wage of $2.70, plus she receives tips. On Monday, she works 8 hours and receives tip money, t. Write an equation showing what Sharice makes on Monday, y.

7. Jenelle buys x shares of stock in a company at $27.80 per share. She later sells the shares at $41.29 per share. Write an equation to show how much money, m, Jenelle has made.

8.6   Going Deeper into the Introduction to Algebra (DOK 3)

A math club made up a code to show the values of 10 variables. They used the code to balance their checkbooks. Anthony made a mistake in his checkbook balance. Unfortunately, errors carry forward making all the balances thereafter, wrong.

1. **First, change all of the amounts into numbers. Then, find the errors and correct them. (DOK 3)**

The code:

A	B	C	D	H	J	K	L	M	P
0	1	2	3	4	5	6	7	8	9

Anthony's Check Register

Date	Name	Withdrawal	Deposit	Balance
8/1	Birthday Money		$CJ.AA	$CJ.AA
8/2	Lee's Pizzeria	$BH.JK		$BA.HH
8/5	Cleaning Garage Money		$CJ.AA	$DJ.HH
8/7	Hanson's Dept. Store	$DB.AP		$J.DH
8/9	School Hot Lunch Ticket	$B.LJ		$D.JP
8/15	Cleaning Gutters Money		$CJ.AA	$CM.JP
8/17	Lee's Pizzeria	$BK.LC		$BB.ML
8/20	Jay's Drugstore	$K.BH		$J.LD
8/21	Washed Dad's Car		$BA.JA	$BK.CD
8/22	Lee's Pizzeria	$BC.KL		$D.JK
8/24	Washed Mom's Car		$BA.JA	$BH.AK

The math club used another code box to give each other practice problems for the upcoming math competition. First find the values of all the variables. Then, find the value of x in each problem. Be sure to do the work within the parenthesis first, then solve. (DOK 3)

The Code for Math Problems:

s	r	t	g	j	w	v	m	p	z
0	1	2	3	4	5	6	7	8	9

2. $t + w + (g \times t) = x$

3. $(z \times g) - (p \div t) = x$

4. $j^2 - g^2 = x$

5. $m \times t - s = x$

6. $(v + v) - (v - g) = x$

7. $r + s + w + m = x$

Follow the directions for the sales problem below. (DOK 3)

12. Jason bought three reams of paper, r, that was on sale at an office supply store. The tax was $0.72. Jason used 3 coupons for $0.50 each. The coupons were applied **after** the tax was calculated.

    Part 1:   Write an algebraic expression that models this scenario.

    Part 2:   If the tax rate is 6%, and Jason paid a total of $11.22, how much did each ream of paper cost?

    Part 3:   Explain the importance of following the order of operations in calculating the total cost of the paper.

Chapter 8 Review

Solve the following problems using $x = 2$. (DOK 1)

1. $2x - 5 =$

2. $\dfrac{18}{x} =$

3. $x^2 - 4 =$

4. $\dfrac{x^3 + 3}{2} =$

5. $11 - 3x =$

6. $x + 20 =$

7. $-4x + 2 =$

8. $8 - x =$

9. $5x - 2 =$

Solve the following problems. Let $w = -\frac{1}{2}$, $y = 2$, $z = 3$. (DOK 2)

10. $4w - y =$

11. $wyz + 3 =$

12. $z - 4w =$

13. $\dfrac{2z + 1}{wz} =$

14. $\dfrac{8w}{y} + \dfrac{z}{w} =$

15. $30 - 5yz =$

16. $-3y + 2 =$

17. $5w - (yw) =$

18. $8y - 3z =$

Write out the algebraic expression given in each word problem. (DOK 2)

19. three less than the sum of x and 10

20. triple Amy's age, a

21. the number of bacteria, b, doubled

22. five less than the product of 4 and y

23. half of a number, n, less 12

24. the quotient of a number, x, and 7

For questions 25–29, write an equation to match each problem. (DOK 3)

25. Calista earns \$300 per week for a 40 hour work week plus \$12.32 per hour for each hour of overtime after 40 hours. Write an equation that would be used to determine her weekly wages where w is her wages, and v is the number of overtime hours worked.

26. Daniel purchased a 1 year CD, for c dollars, from a bank. He bought it at an annual interest rate of 7%. After 1 year, Daniel cashes in the CD. Write an equation that would determine the total amount it is worth, x.

27. Omar is a salesman. He earns an hourly wage of \$12.00 per hour plus he receives a commission of 5% on the sales he makes. Write an equation which would be used to determine his weekly salary, w, where x is the number of hours worked, and y is the amount of sales for the week.

28. Juan sold a boat that he bought 5 years ago. He sold it for 70% less than he originally paid for it. If the original cost is b, write an expression that shows how much he sold the boat for.

29. Toshi is going to get a 5% raise after he works at his job for 1 year. If s represents his starting salary, write an expression that shows how much he will make after his raise.

Chapter 8 Test

1 Tom earns $400 per week before taxes are taken out. His employer takes out a total of 31% for state, federal, and Social Security taxes. Which expression below will help Tom figure his net pay?

  A $400 - 0.31$

  B $400 + 0.31$

  C $400 - 0.31\,(400)$

  D $400 + 0.31(400)$

(DOK 2)

2 Simplify the following expression using $x = 2$ and $y = 5$.

$$3x + 4y - 1$$

  A 25

  B 13

  C 22

  D 10

(DOK 2)

3 A plumber charges $37.00 per hour plus a $15.00 service charge. If a represents his total charges in dollars, and b represents the number of hours worked, which formula below could the plumber use to calculate his total charges?

  A $a = 37 + 15b$

  B $a = 37 + 15 + b$

  C $a = (37)\,(15) + b$

  D $a = 37b + 15$

(DOK 3)

4 Which expression is a number divided by the sum of nine and two.

  A $\frac{x}{9} + 2$

  B $\frac{9+2}{x}$

  C $\frac{x}{9+2}$

  D $\frac{x+2}{9}$

(DOK 2)

5 In 2011, Bell computers announced to its sales force to expect a 2.4% price increase on all computer equipment in the year 2012. A certain sales representative wanted to see how much the increase would be on a computer, c, that sold for $2,600 in 2011. Which expression below will help him find the cost of the computer in the year 2012?

  A $0.24\,(2{,}600)$

  B $2{,}600 + 0.024\,(2{,}600)$

  C $0.24\,(2{,}600) + 2{,}600$

  D $0.024\,(2{,}600) - 2{,}600$

(DOK 2)

Use the table of values below to answer questions 6–7.

a	b	c	d	e
2	3	4	5	8

6 Solve for x: $(e \div a) + (b + c + d) = x$

  A 16

  B 22

  C 18

  D 17

(DOK 3)

7 Solve for x: $(d \times c) + (e \times b) = x$

  A 33

  B 31

  C 20

  D 44

(DOK 3)

8 Solve for x: $b + a - (d \times e) = x$

  A 35

  B -35

  C -8

  D -34

(DOK 3)

Chapter 9
Equations and Inequalities

This chapter covers the following CC Grade 7 standards:

	Content Standards
Expressions and Equations	7.EE.1, 7.EE.3, 7.EE.4

9.1   Two-Step Equations (DOK 2)

Rule 1: Put all the integers on one side of the equal sign.

This can be done by adding or subtracting the integers from both sides.

If the integer is positive, subtract the same integer from both sides of the equation.

If the integer is negative, add the same integer to both sides of the equation.

Rule 2: Divide or multiply both sides by the coefficient of the variable.

(Example: $3x$ is equal to $3 \times x$. Because this is multiplication, you must do the opposite. Divide both sides of the equation by 3 to isolate the number from the variable, x.)

Refer to the "About AR" on page xi!

Example 1:   Solve $3x - 5 = 10$ for x.

$$\begin{array}{rcl} 3x \quad -5 & = & 10 \\ +5 & = & +5 \\ \hline 3x & = & 15 \end{array}$$ First, add 5 to both sides.

$$\frac{3x}{3} = \frac{15}{3}$$ Next, divide both sides by 3 to get the answer.

$$x = 5$$

To check, replace x with 5 in the original problem.

$$3(5) - 5 = 10$$

$$15 - 5 = 10$$

$x = 5$ makes this a true sentence.

Solve the following two step equations. Double check your answer by replacing the variable with your solution to see if this will make the sentence true. (DOK 2)

1. $2s + 5 = 9$

2. $7c + 6 = 13$

3. $3x + 2 = 14$

4. $2 + 5x = 12$

5. $3k - 2 = 7$

6. $4r - 6 = 10$

7. $3n + 6 = 9$

8. $7x - 1 = 62$

9. $5y + 9 = 79$

10. $7y + 7 = 14$

11. $4x + 4 = 8$

12. $4x - 4 = 8$

13. $2w - 1 = 1$

14. $5w + 8 = 53$

15. $2x - 10 = 18$

16. $19m - 12 = 26$

17. $110y + 10 = 340$

18. $6j - 6 = 18$

19. $20r + 3 = 103$

20. $x + 1 = 1$

21. $9y - 2 = 25$

22. $11x + 12 = 34$

23. $4b - 2 = 42$

24. $3x + 7 = 43$

9.2   Two-Step Equations with Rational Numbers (DOK 2)

The rules for solving two-step equations can be applied to solving two-step equations with rational numbers with only some minor changes:

Rule 1: Put all the rational numbers on one side of the equal sign.

This can be done by adding or subtracting the rational numbers from both sides.

If the rational number is positive, subtract the same rational number from both sides of the equation.

If the rational number is negative, add the same rational number to both sides of the equation.

Rule 2: Divide or multiply both sides by the coefficient of the variable.

(Example: $\frac{1}{3}x$ is equal to $\frac{1}{3} \times x$. Because this is multiplication, you must do the opposite to both sides of the equation to isolate the number from the variable, x. In other words, you must divide. Remember that dividing fractions is also the same as multiplying by the reciprocal.)

Example 2:   Solve for x: $5x - 2 = 6.3$

Step 1:   Add 2 to both sides of the equation:

$$\begin{array}{rrcr} 5x & - \ 2 & = & 6.3 \\ & + \ 2 & & +2 \\ \hline 5x & & = & 8.3 \end{array}$$

Step 2:   Divide both sides of the equation by 5: $\dfrac{5x}{5} = \dfrac{8.3}{5}$

Answer: $x = 1.66$

Solve the following two step equations with rational numbers. Double check your answer by replacing the variable with your solution to see if this will make the sentence true. (DOK 2)

1. $\frac{2}{3}x + 8 = 20$

2. $1.25x - 3.2 = 9.55$

3. $\frac{1}{4} + 1\frac{1}{2}x = \frac{3}{8}$

4. $0.25 + 1.11x = 5.8$

5. $\frac{1}{2}x + \frac{1}{2} = \frac{1}{2}$

6. $\frac{1}{2}x - \frac{1}{2} = \frac{1}{2}$

7. $9x - 6 = 7.5$

8. $3.3x + 9.8 = 19.04$

9. $\frac{2}{5} + \frac{3}{10}x = \frac{5}{9}$

10. $9.3 + 5.6x = 26.1$

11. $15.82x - 13.5 = 6.275$

12. $1\frac{1}{5}x + \frac{2}{3} = 1\frac{2}{3}$

13. $18x - 1\frac{5}{9} = -1$

14. $5.9x - 1.026 = 17.5$

15. $\frac{1}{8}x + \frac{3}{5} = 8\frac{3}{5}$

16. $0.2x - 5.6 = -3.2$

9.3   Combining Like Terms (DOK 1)

In algebra problems, separate **terms** by $+$ and $-$ signs. The expression $5x - 4 - 3x + 7$ has 4 terms: $5x$, 4, $3x$, and 7. Terms having the same variable can be combined (added or subtracted) to simplify the expression. $5x - 4 - 3x + 7$ simplifies to $2x + 3$.

$$5x - 3x \quad - 4 + 7 \ = 2x + 3$$

Simplify the following expressions.

1. $4x + 8x =$

2. $3y - 5y + 8 =$

3. $2 - 2x + 3 =$

4. $9a - 14 - a =$

5. $7w + 3w + 2 =$

6. $-2x + x + 8x =$

7. $w - 4 + 5w =$

8. $5 - 10t + 3 - 7t =$

9. $-2 + x - 4x + 8 =$

10. $12b + 10 + 6b =$

11. $9h - h + 2 - 4 =$

12. $-3k + 8 - 4k =$

13. $2a + 10a - 2 + a =$

14. $2 + 9c - 12 =$

15. $-d + 4 + 3d - 1 =$

16. $-8 + 3h + 5 - h =$

17. $10x - 5x + 8 =$

18. $11 + 4z + z - 6 =$

19. $11 + 7y - y - 2 =$

20. $14p - 2 + p =$

21. $15m + 2 - m + 3 =$

9.4   Removing Parentheses (DOK 2)

In this chapter, you will use the distributive property to remove parentheses in problems with a variable.

Example 3: $2(a + 6)$

You multiply 2 by each term inside the parentheses. $2 \times a = 2a$ and $2 \times 6 = 12$. The 12 is a positive number so use a plus sign between the terms in the answer.

$2(a + 6) = 2a + 12$

Example 4: $7(2b - 5)$ $7 \times 2b = 14b$ and $7 \times -5 = -35$

$7(2b - 5) = 14b - 35$

Example 5: $4(-5c + 2)$

The first term inside the parentheses could be negative. Multiply in exactly the same way as the examples above. $4 \times (-5c) = -20c$ and $4 \times 2 = 8$

$4(-5c + 2) = -20c + 8$

Remove the parentheses in the problems below. (DOK 2)

1. $3(n+2)$

2. $7(2g-11)$

3. $2(6z-2)$

4. $7(-y-3)$

5. $5(-3k+9)$

6. $9(d-4)$

7. $3(-4x+8)$

8. $8(4+7p)$

9. $12(-4w-4)$

10. $2(9x+2)$

11. $3(2-y)$

12. $10(c-7)$

13. $5(-3t+1)$

14. $4(4y+3)$

15. $10(b+5)$

16. $6(2a+7)$

17. $9(2b-3)$

18. $7(-9x-8)$

19. $7(8-9v)$

20. $3(3c+8)$

21. $6(2x-4)$

22. $8(y+6)$

23. $2(7t+5)$

24. $4(9-g)$

The number in front of the parentheses can also be negative. Remove these parentheses the same way.

Example 6: $-2(b-4)$

First, multiply $-2 \times b = -2b$

Second, multiply $-2 \times -4 = 8$

Copy the two products. The second product is a positive number, so put a plus sign between the terms in the answer.

$-2(b-4) = -2b+8$

Remove the parentheses in the following problems. (DOK 2)

1. $-2(x+3)$

2. $-4(2-y)$

3. $-9(2b-5)$

4. $-3(8c+4)$

5. $-10(-w-7)$

6. $-2(4x-3)$

7. $-5(-z+3)$

8. $-2(7p+9)$

9. $-6(t-2)$

10. $-18(2w+5)$

11. $-4(9-8p)$

12. $-3(-k-4)$

13. $-7(7b-11)$

14. $-9(-5t-12)$

15. $-8(-v+20)$

16. $-5(-x-4)$

17. $-12(4y+7)$

18. $-3(-c+8)$

19. $-4(-2t-7)$

20. $-8(7z-8)$

21. $-36(y-2)$

22. $-200(a+2)$

23. $-4(-x-9)$

24. $-20(-2b+2)$

9.5   Solving Two-Step Algebra Word Problems (DOK 2)

Just like solving any word problem, the key is to READ CAREFULLY.

Example 7:   A taxi costs a flat rate of $3.00 and then an additional $0.75 per mile. Let d equal the miles traveled. Find the miles traveled, d, for a total fare of $14.25.

Step 1:   Find what is known and unknown.
Known: $3.00 is part of the fare; $0.75 per mile is part of the fare; $14.25 is the total fare.
Unknown: The number of miles traveled, d.

Step 2:   Set up the equation and solve.

$$\begin{array}{rcl} \$3.00 + \$0.75d & = & \$14.25 \\ -\ \$3.00 & = & -\ \$3.00 \\ \hline \dfrac{0.75d}{0.75} & = & \dfrac{\$11.25}{0.75} \\ d & = & 15 \end{array}$$

Subtract $3.00 from both sides.

Divide both sides by 0.75.

Answer: 15 miles were traveled.

Solve the two-step equation word problems. (DOK 2)

1. Mrs. Beasley bought a number of juice boxes at a cost of $0.30 each and a bag of chips that cost $1.19. The total cost of her purchases was $2.99. Write an equation and solve to find out how many juice boxes Mrs. Beasley bought.

2. Cathy had saved $335.00. She spent $15.00 each week to pay for guitar lessons. She now has $155.00. Write an equation and solve to find out how many weekly lessons Cathy had.

3. Paula sold 45 packages of wrapping paper for the fall fund-raiser. This is three less than twice the number of packages that she sold last year. Write and solve an equation to find how many packages Paula sold last year.

4. Movie tickets for two children cost $24 minus $12. Find the cost of a child's ticket.

5. The Mogul Runners Ski Club planned a trip to Park City. Of the total number of members, 11 signed up to go. If this is 25% of the club, how many total members does the ski club have?

6. Five children in the Woods family have decided to purchase an anniversary gift for their parents. They estimate the price of the gift to be $250. If they divide the cost of the gift evenly among them, how much does each of the children have to pay?

9.6   Graphing Inequalities (DOK 2)

An inequality is a sentence that contains a $<$, $>$, $\leq$, or $\geq$ sign. Look at the following graphs of inequalities on a number line. Graphing an inequality shows all the numbers that make the inequality true.

$x < 3$ is read "x is less than 3" since it is $<$ and not $\leq$.

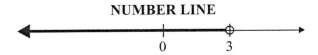

NUMBER LINE

$x \leq 5$ is read "x is less than or equal to 5" since the inequality is $\leq$. The endpoint is filled in. The graph uses a **closed** circle because the number 5 is <u>included</u> in the graph.

$x > -2$ is read "x is greater than -2."

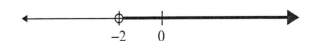

$x \geq 1$ is read "x is greater than or equal to 1."

There can be more than one inequality sign. These are called **compound inequalities**. For example:

$-2 \leq x < 4$ is read "-2 is less than or equal to x and x is less than 4."

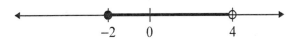

$x < 1$ or $x \geq 4$ is read "x is less than 1 or x is greater than or equal to 4."

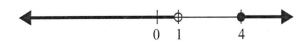

Graph the solution sets of the following inequalities. (DOK 2)

1. $x > 8$

2. $x \leq 5$

3. $-5 < x < 1$

4. $x > 7$

5. $1 \leq x < 4$

6. $x < -2$ or $x > 1$

7. $x \geq 10$

8. $x < 4$

9. $x \leq 3$ or $x \geq 5$

10. $x < -1$ or $x > 1$

Give the inequality represented by each of the following number lines. (DOK 2)

11.  _____

12. _____

13. _____

14.  _____

15.  _____

16.  _____

17. 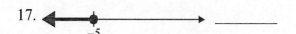 _____

18. _____

9.7   Solving Inequalities by Addition and Subtraction (DOK 2)

Solving inequalities is similar to solving equations.

Example 8:   Solve and graph the solution set for $x - 2 \leq 5$.

Step 1:   Add 2 to both sides of the inequality so the variable will be by itself.

$$x - 2 \leq 5$$
$$\underline{+2 \ +2}$$
$$x \leq 7$$

Step 2:   Graph the solution set for the inequality.

Solve and graph the solution set for the following inequalities. (DOK 2)

1. $x + 5 > 3$

2. $x - 10 < 5$

3. $x - 2 \leq 1$

4. $9 + x \geq 7$

5. $x - 4 > -2$

6. $x + 11 \leq 20$

7. $x - 3 < -12$

8. $x + 6 \geq -3$

9. $x + 12 \leq 8$

10. $15 + x > 5$

11. $x - 6 < -2$

12. $x + 7 \geq 4$

13. $14 + x \leq 8$

14. $x - 8 > 24$

15. $x + 1 \leq 12$

16. $11 + x \geq 11$

17. $x - 3 < 17$

18. $x + 9 > -4$

19. $x + 6 \leq 14$

20. $x - 8 \geq 19$

9.8   Solving Inequalities by Multiplication and Division (DOK 2)

If you multiply or divide both sides of an inequality by a **positive** number, the inequality symbol stays the same. However, if you multiply or divide both sides of an inequality <u>by a</u> **negative** number, **you must reverse the direction of the inequality symbol.**

Example 9:   Solve and graph the solution set for $4x \le 20$.

Step 1:   Divide both sides of the inequality by 4. $\dfrac{\cancel{4}x}{\cancel{4}} \le \dfrac{\cancel{20}}{\cancel{4}}$

Step 2:   Graph the solution. $x \le 5$

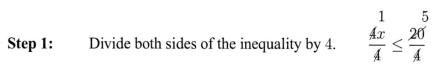

Example 10:   Solve and graph the solution set for $6 > -\dfrac{x}{3}$.

Step 1:   Multiply both sides by -3 and **reverse the direction of the inequality symbol.**

$$(-3) \times 6 < \frac{x}{\cancel{-3}} \times \cancel{-3}$$

Step 2:   Graph the solution. $-18 < x$
$x > -18$

Solve and graph the following inequalities. (DOK 2)

1. $\dfrac{x}{5} > 4$

2. $2x \le 24$

3. $-6x \ge 36$

4. $\dfrac{x}{10} > -2$

5. $-\dfrac{x}{4} > 8$

6. $-7x \le -49$

7. $-3x > 18$

8. $-\dfrac{x}{7} \ge 9$

9. $9x \le 54$

10. $\dfrac{x}{8} > 1$

11. $-\dfrac{x}{9} \le 3$

12. $-4x < -12$

13. $-\dfrac{x}{2} \ge -20$

14. $10x \le 30$

15. $\dfrac{x}{12} > -4$

16. $-6x < 24$

9.9   Two-Step Inequalities (DOK 2)

Remember that adding and subtracting with inequalities follow the same rules as equations. When you multiply or divide both sides of an inequality by the same positive number, the rules are also the same as for equations. However, when you multiply or divide both sides of an inequality by a **negative** number, you must **reverse** the inequality symbol.

Example 11: $-x > 4$
$(-1)(-x) < (-1)(4)$
$x < -4$

Example 12: $-4x < 2$

$$\frac{-4x}{-4} > \frac{2}{-4}$$

$$x > -\frac{1}{2}$$

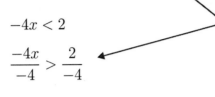

Reverse the symbol when you multiply or divide by a negative number.

When solving multi-step inequalities, first add and subtract to isolate the term with the variable. Then multiply and divide.

Example 13: $-2x - 8 > 1$

Step 1:   Add 8 to both sides.

$-2x - 8 + 8 > 1 + 8$

$-2x > 9$

Step 2:   Divide by -2. Remember to change the direction of the inequality sign.

$$\frac{-2x}{-2} < \frac{9}{-2}$$

Answer: $x < -\dfrac{9}{2}$

Solve each of the following inequalities. (DOK 2)

1. $8 - 3x \leq 1$

2. $2x - 4 \geq 2$

3. $\frac{1}{4}b - 3 > 5$

4. $8 + 3y > -4$

5. $5a + 5 < -6$

6. $-\frac{x}{3} > 18$

7. $3x - 7 \geq 2$

8. $6x - 3 \leq 4$

9. $3(a - 2) > -2$

10. $-\frac{2x}{6} \leq 4$

11. $9b + 5 < 8$

12. $4x - 9 \leq 9$

13. $5x + 3 \leq -2$

14. $3y - 4 > 6$

15. $3 - 2y \leq -25$

16. $-4c + 8 \leq 10$

17. $-\frac{1}{3}x + 5 > 9$

18. $\frac{1}{6}y - 3 \leq 2$

19. $-3x + 2 > 8$

20. $\frac{y}{3} - 2 \geq 5$

21. $6 + 8c < -2$

22. $4 - \frac{a}{2} > 3$

23. $6 + 4b \leq -3$

24. $-\frac{1}{3}x + 2 > 4$

Graph the answers from problems 10–24 above on the number lines below. (DOK 2)

10. ⟵—————————⟶

11. ⟵—————————⟶

12. ⟵—————————⟶

13. ⟵—————————⟶

14. ⟵—————————⟶

15. ⟵—————————⟶

16. ⟵—————————⟶

17. ⟵—————————⟶

18. ⟵—————————⟶

19. ⟵—————————⟶

20. ⟵—————————⟶

21. ⟵—————————⟶

22. ⟵—————————⟶

23. ⟵—————————⟶

24. ⟵—————————⟶

9.10   Solving Inequality Word Problems (DOK 2)

Write an inequality for the word problems below, then solve. (DOK 2)

1. Jacob weighs 50 pounds more than his younger brother, Jared. Jared weighs 46 pounds. Let x represent Jacob's weight. How much does Jacob weigh?

2. Alice is 16 years old. If you add together Alice's age and her cousin Emily's age, x, together, you will get more than 31 years. How old is Emily?

3. Maria scored more than 4 times as many points as Jennifer, who scored 6 points. How many points, x, did Maria score?

4. This month's sales of $85,000$ are greater than $\frac{1}{2}$ of the sales of March, represented by the letter s. What was March's sales?

5. The Furry Friends Animal Shelter had less than 27 dogs at the beginning of the month. At the end of the month, they had 14 dogs. Let x represent how many dogs were adopted. How many dogs were adopted? (Assume they did not take in any extra dogs.)

6. Pine City (x) is less than or equal to 20 miles from Sun Valley, depending on the route you take. How many miles is Pine City from Sun Valley?

7. Molly's age, x, plus Betty's age, 11, is less than 23. What is Molly's age?

8. There is more than twice as much rain in November, x, as there is in August, which had 4 inches. How much rain was in November?

9. Mark had to read more than 200 pages in 5 days. How many pages, x, must Mark read each day to complete the assignment?

10. If the temperature, x, goes up 6 degrees, it will be hotter than 100 degrees. What is the temperature outside?

9.11   Going Deeper into Equations (DOK 3)

Read the problems, write an equation to fit the problem, combine like terms, and solve. Show your work for each step. (DOK 3)

1. Abby bought 4 times as many music downloads as her brother, Andy, bought on Saturday. Andy bought x music downloads. Their sister Anna bought 2 times as many music downloads as her brother. Their father bought 3 times as many plus 2 music downloads as his son. Their mother bought 1 less music download as her son. In all, Abby and her family bought 23 music downloads. How many music downloads did each person buy?

2. The Drake family spent the day at the beach digging for clams. Aaron found x clams. His sister, Molly, found 2 times as many clams minus 3 than Aaron found. Skyler found 4 clams less than Molly. Mr. Drake found 2 times plus 3 as many clams as Aaron. Mrs. Drake found 2 times as many clams as Molly. The Drake family found 64 clams in all. How many clams did each family member find?

3. Maddox has 4 ants left in his ant farm. Ant Enna moved x bread crumbs to the nest. Ant Body moved twice as many less 2 bread crumbs than Ant Enna to the nest. Ant Cee moved three times as many bread crumbs to the nest as Ant Body. Ant Mable moved three times plus 3 bread crumbs as Ant Enna. The ants moved 55 bread crumbs in all. How many bread crumbs did each ant move to the nest?

4. A new package has 30 cookies in it. Elwood took x number of cookies, James took $\frac{1}{2}$ as many as Elwood, Arthur took twice as many as Elwood, and Phillip took $\frac{3}{4}$ as many as Arthur. That emptied the package. How many cookies did each boy get?

5. A factory is doing end-of-year inventory. Hank, Frank, Patrick, and Brent are counting the remaining model #LG3039 in the four boxes left on the shelves. Hank counted x number in his box. Frank counted 4 times as many less 10 in his box. Patrick counted twice as many as Frank in his box. And Brent counted 3 times plus 5 as many as Hank. They counted a total of 87 model #LG3039. How many did each man count?

6. Three friends are in a friendly race to finish their math homework. Madison has x number of problems left to do. Bella has double less 3 as many problems as Madison has left to do. Emily has 7 less than 2 times the number of problems as Madison has left to do. The three girls have 10 problems left to do. How many problems does each girl have left to do?

7. Four friends timed how long it took to do their homework. Jasper took x number of minutes. Hannah took 3 times as long less 8 minutes than Jasper. Mark took 2 times plus 5 minutes as Jasper took to do his homework. Olivia took 2 times plus 9 minutes as Jasper to do her homework. In all, the friends spent 78 minutes doing their homework. How long did it take each friend to do their homework?

Chapter 9 Review

Solve each of the following equations. (DOK 2)

1. $3a - 8 = 16$

2. $-16 + 2w = 54$

3. $6 + \dfrac{x}{3} = -2$

4. $\dfrac{c}{12} - 3 = 4$

5. $\dfrac{y - 7}{3} = 2$

6. $\dfrac{b + 9}{8} = -4$

Solve. (DOK 2)

7. $6d - 10 = 14$

8. $-16x - 34 = 14$

9. $10w - 2 = -22$

Solve each of the following equations and inequalities. (DOK 2)

10. $3b - 12 = -20$

11. $17x + 15 = -16$

12. $43c - 32 = -35$

13. $-3x - 6 = -15$

14. $4b - 6 = -5$

15. $2a - 7 = -26$

16. $2x + 5 \geq 4$

17. $8x + 10 = 12$

18. $4x + 12 < -14$

19. $-3x + 4 < -8$

20. $8x + 2 > 0$

21. $-\dfrac{y}{3} > 12$

22. $-\dfrac{3}{4}x \leq 12$

23. $6x + 3 \geq -6$

24. $n + 20 = -40$

25. $\dfrac{t}{3} + 2 > 7$

Graph the solution sets of the following inequalities. (DOK 2)

26. $x \leq -5$

27. $x > 12$

28. $x < -4$

29. $x \geq 8$

Give the inequality represented by each of the following number lines. (DOK 2)

30.

31.

32.

33.

Simplify the following expressions by combining like terms. (DOK 1)

34. $-4a + 14 + 7a - 9$ 35. $7 + 3z - 8 - 3z$ 36. $-5 - 7x - 3 - 7x$

Simplify the following expression by removing parentheses. (DOK 2)

37. $2(-4x + 5)$ 39. $4(8 - 6b)$ 41. $-2(4c - 2)$

38. $15(2y + 3)$ 40. $-5(-4 + 3a)$ 42. $-3(7y - 4)$

Solve the inequality word problems. (DOK 2)

43. Mrs. Watson is 28 years older than her youngest daughter Sarah's age, x. Mrs. Watson is less than 54 years old. How old is Sarah?

44. Mr. Andrews paid less than 50 cents more per gallon of gas, x, this week than last week, when he paid $2.50 per gallon. How much was gas this week?

45. Jim takes great pride in decorating his club's float for the homecoming parade for his high school. With the $5,000 he has to spend, Jim bought 5,000 carnations at $0.25 each, 4,000 tulips at $0.50 each, and 300 irises at $0.90 each. Write an inequality which describes how many roses, r, Jim can buy if roses cost $0.80 each.

Read the problems, write an equation to fit the problem, combine like terms, and solve. Show your work for each step. (DOK 3)

46. Don collected x number of agates (a kind of rock) on the shores of Lake Superior on his summer vacation. His brother, Bryan, collected twice less 9 agates than Don. Their sister Ann collected 3 times plus 2 agates than Don. Their father collected twice as many agates as Ann. In all, they collected 981 agates. How many agates did each family member find?

47. A class is collecting aluminum cans to earn money to help pay for a trip to the astronomy museum. Ethan collected x number of cans. Mike collected 4 times minus 7 cans than Ethan. Dana collected 3 times plus 10 cans than Ethan. Alexis collected twice plus 17 cans than Ethan. In all the four students collected 420 cans. How many cans did each of the four students collect?

Chapter 9 Test

1 Solve: $3x - 7 = 14$

  A $x = 21$
  B $x = 3$
  C $x = 14$
  D $x = 7$

(DOK 2)

2 $\dfrac{x}{4} + 2 = 66$

  A $x = 16$
  B $x = 256$
  C $x = 64$
  D $x = 17$

(DOK 2)

3 Solve: $3y + 2 = -5$

  A $y = 7$
  B $y = -7$
  C $y = 9$
  D $y = -\frac{7}{3}$

(DOK 2)

4 Solve: $\frac{1}{4}w + 4 < 16$

  A $w < 48$
  B $w < 4$
  C $w > 12$
  D $w < 12$

(DOK 2)

5 Solve: $15x + 17 = -43$

  A $x = -4$
  B $x = 4$
  C $x = -26$
  D $x = 26$

(DOK 2)

6 Solve: $0.2x + 27 = 82$

  A $x = 15$
  B $x = 21.8$
  C $x = 55$
  D $x = 275$

(DOK 2)

7 Which inequality does the number line represent?

  A $x \leq -1$
  B $x < -1$
  C $x > -1$
  D $x \geq -1$

(DOK 2)

8 Solve for x: $\dfrac{x - 8}{2} = 7$

  A 22
  B 6
  C 28
  D 8

(DOK 2)

9 Solve for w: $4w + 12 = -26$

  A 3.5
  B -3.5
  C -9.5
  D 9.5

(DOK 2)

10 Solve for d: $-6d + 16 = -20$

  A 6
  B -6
  C -2
  D -5

(DOK 2)

11 Solve for g: $-5g - 25 \geq 10$

   A $g \geq -7$

   B $g \geq 7$

   C $g \leq 7$

   D $g \leq -7$

(DOK 2)

12 Solve for x: $2x - 2 = 6$

   A 3

   B 4

   C 5

   D 6

(DOK 2)

13 Jeremy earned x number of dollars mowing lawns in the month of June. His brother, Matthew, earned twice plus \$4 of what Jeremy earned. Their cousin, Dan, earned 3 times less \$10 of what Jeremy earned. The three boys earned \$264. How much did each boy earn?

   A Jeremy earned \$45
      Matthew earned \$90
      Dan earned \$125

   B Jeremy earned \$45
      Matthew earned \$904
      Dan earned \$135

   C Jeremy earned \$45
      Matthew earned \$94
      Dan earned \$125

   D Jeremy earned \$45
      Matthew earned \$94
      Dan earned \$135

(DOK 3)

14 Simplify: $14g + 12 + 2g - 7 =$

   A $16g + 5$

   B $14g + 19$

   C $12g + 19$

   D $12g + 5$

(DOK 1)

15 Remove the parentheses: $-6(2x + 9)$

   A $-12x + 54$

   B $-12x - 54$

   C $12x + 3$

   D $2x + 3$

(DOK 2)

16 Which inequality represents the following? If Janelle adds 8 more carrots to a serving plate, there will be more than 22 carrots on the plate.

   A $x - 8 = 22$

   B $x + 8 > 22$

   C $x + 8 < 22$

   D $x + 8 = 22$

(DOK 2)

17 Janet and Artie want to play tug-of-war. Artie pulls with 200 pounds of force while Janet pulls with 60 pounds of force. In order to make this a fair contest, Janet enlists the help of her friends Trudi, Sherri, and Bridget who pull with 20, 25, and 55 pounds respectively. Write an inequality describing the minimum amount Janet's fourth friend, Tommy, must pull to beat Artie.

   A $x > 40$ pounds of force

   B $x < 40$ pounds of force

   C $x > 100$ pounds of force

   D $x < 100$ pounds of force

(DOK 3)

Chapter 10
Angles

This chapter covers the following CC Grade 7 standard:

	Content Standard
Geometry	7.G.5

10.1 Angles (DOK 1)

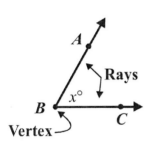

Angles are made up of two rays with a common endpoint, called the vertex. Rays are named starting with the endpoint and then another point on the ray. Ray $\overrightarrow{BA}$ and ray $\overrightarrow{BC}$ share a common endpoint.

Angles are usually named by three capital letters. The middle letter names the vertex. The angle to the left can be named $\angle ABC$ or $\angle CBA$. Sometimes an angle can be named by a lower case letter between the sides, $\angle x$, or by the vertex alone, $\angle B$.

A protractor, shown below, is used to measure angles. The protractor is divided evenly into a half circle of 180 degrees (180°). When the middle of the bottom of the protractor is placed on the vertex, and one of the rays of the angle is lined up with 0°, the other ray of the angle crosses the protractor at the measure of the angle. The angle below has the ray pointing left lined up with 0° (the outside numbers), and the other ray of the angle crosses the protractor at 55°. The angle measures 55°.

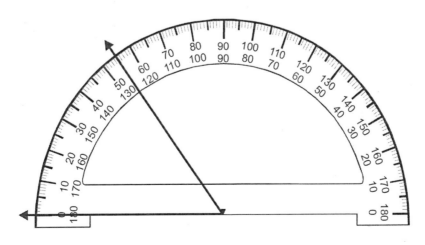

10.2   Types of Angles (DOK 1)

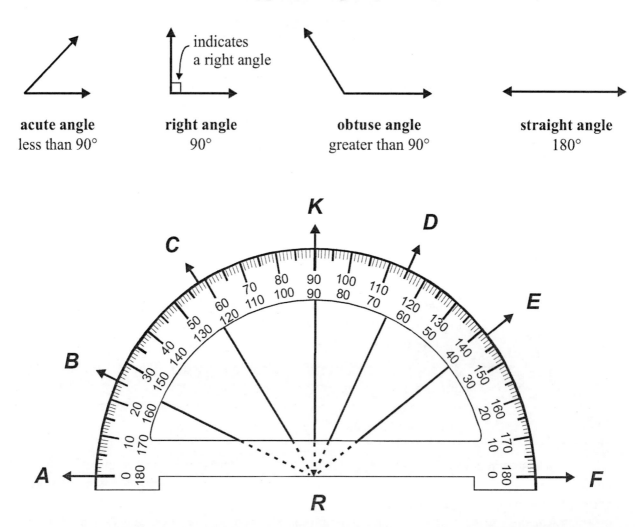

acute angle
less than 90°

right angle
90°

indicates
a right angle

obtuse angle
greater than 90°

straight angle
180°

Using the protractor above, find the measure of the following angles. Then, tell what type of angle it is: acute, right , obtuse, or straight. (DOK 1)

		Measure	**Type of Angle**
1.	What is the measure of ∠ARF?	_____	_____
2.	What is the measure of ∠CRF?	_____	_____
3.	What is the measure of ∠BRF?	_____	_____
4.	What is the measure of ∠ERF?	_____	_____
5.	What is the measure of ∠ARB?	_____	_____
6.	What is the measure of ∠KRA?	_____	_____
7.	What is the measure of ∠CRA?	_____	_____
8.	What is the measure of ∠DRF?	_____	_____
9.	What is the measure of ∠ARD?	_____	_____
10.	What is the measure of ∠FRK?	_____	_____

10.3 Adjacent Angles (DOK 1)

Adjacent angles are two angles that have the same vertex and share one ray. They do not share space inside the angles.

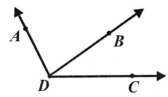

$\angle ADB$ is **adjacent** to $\angle BDC$. However, $\angle ADB$ is **not adjacent** to $\angle ADC$ because adjacent angles do not share any space inside the angle

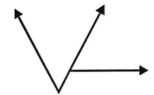

These two angles are **not adjacent**. They share a common ray but do not share the same vertex.

For each diagram below, name the adjacent angle. (DOK 1)

1.

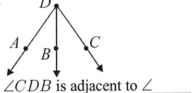

$\angle CDB$ is adjacent to $\angle$_____

5.

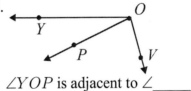

$\angle YOP$ is adjacent to $\angle$_____

2.

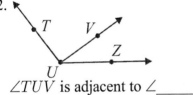

$\angle TUV$ is adjacent to $\angle$_____

6.

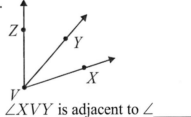

$\angle XVY$ is adjacent to $\angle$_____

3.

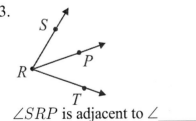

$\angle SRP$ is adjacent to $\angle$_____

7.

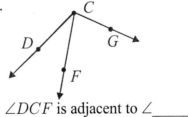

$\angle DCF$ is adjacent to $\angle$_____

4.

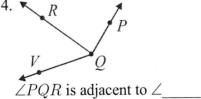

$\angle PQR$ is adjacent to $\angle$_____

8.

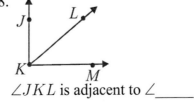

$\angle JKL$ is adjacent to $\angle$_____

10.4   Vertical Angles (DOK 2)

When two lines intersect, two pairs of vertical angles are formed. Vertical angles are not adjacent. Vertical angles are congruent, that is, have the same measure.

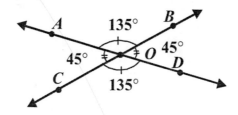

$\angle AOB$ and $\angle COD$ are vertical angles. $\angle AOC$ and $\angle BOD$ are vertical angles.

In the diagram below, name the second angle in each pair of vertical angles. (DOK 2)

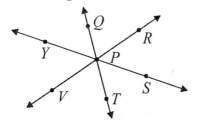

1. $\angle YPV$ _____
2. $\angle QPR$ _____
3. $\angle SPT$ _____

4. $\angle VPT$ _____
5. $\angle RPT$ _____
6. $\angle VPS$ _____

7. $\angle MLN$ _____
8. $\angle KLH$ _____
9. $\angle GLN$ _____

10. $\angle GLM$ _____
11. $\angle KLM$ _____
12. $\angle HLG$ _____

Use the information given to find the measure of each unknown vertical angle. (DOK 2)

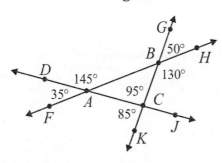

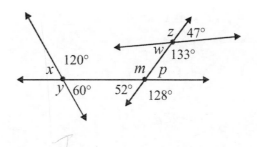

13. $\angle CAF =$ _____

14. $\angle ABC =$ _____

15. $\angle KCJ =$ _____

16. $\angle ABG =$ _____

17. $\angle BCJ =$ _____

18. $\angle CAB =$ _____

19. $\angle x =$ _____

20. $\angle y =$ _____

21. $\angle z =$ _____

22. $\angle w =$ _____

23. $\angle m =$ _____

24. $\angle p =$ _____

10.5 Complementary and Supplementary Angles (DOK 2)

Two angles are **complementary** if the sum of the measures of the angles is 90°.

Two angles are **supplementary** if the sum of the measures of the angles is 180°.

The angles may be adjacent but do not need to be.

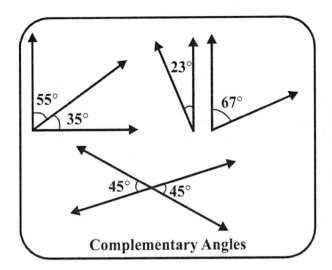

Complementary Angles

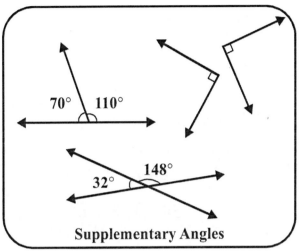

Supplementary Angles

Calculate the measure of each unknown angle. (DOK 2)

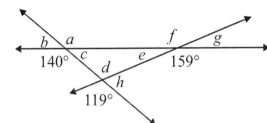

1. $\angle a =$ _____
2. $\angle b =$ _____
3. $\angle c =$ _____
4. $\angle d =$ _____

5. $\angle e =$ _____
6. $\angle f =$ _____
7. $\angle g =$ _____
8. $\angle h =$ _____

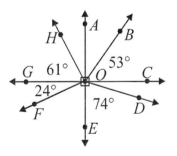

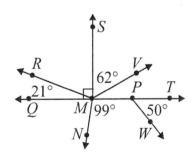

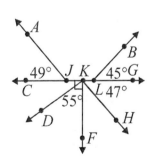

9. $\angle AOB =$ _____

10. $\angle COD =$ _____

11. $\angle EOF =$ _____

12. $\angle AOH =$ _____

13. $\angle RMS =$ _____

14. $\angle VMT =$ _____

15. $\angle QMN =$ _____

16. $\angle WPQ =$ _____

17. $\angle AJK =$ _____

18. $\angle CKD =$ _____

19. $\angle FKH =$ _____

20. $\angle BLC =$ _____

10.6   Finding Angles in Figures (DOK 2)

Unknown angles can be found using facts about supplementary, complementary, and vertical angles.

Other helpful facts: the interior angles of a triangle sum to $180°$
the interior angles of a quadrilateral sum to $360°$

Example 1:   Solve for the unknown angles.

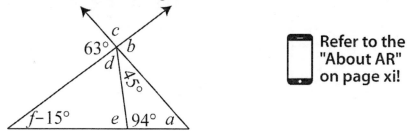

Refer to the "About AR" on page xi!

Step 1:   The sum of interior angles in a triangle is $180°$.
Therefore, $45° + 94° + a = 180°$.
$a = 41°$

Step 2:   Vertical angles are equal in measure, so $b = 63°$.

Step 3: $\angle b$ and $\angle c$ are supplementary angles, so $b + c = 180°$.
Substitute the known value for b and solve for c.
$63° + c = 180°$ $c = 117°$

Step 4:   Using the definition of vertical angles, $c = d + 45°$.
Substitute the known value for c and solve for d.
$117° = d + 45°$ $d = 72°$

Step 5:   Using the definition of supplementary angles, $e + 94° = 180°$.
Solve for e.
$e = 86°$

Step 6:   The sum of interior angles in a triangle is $180°$.
Therefore, $d + e + (f - 15°) = 180°$
Substitute the known value for d and e and solve for f.
$72° + 86° + f - 15° = 180°$ $f = 37°$

Answer: $a = 41°, b = 63°, c = 117°, d = 72°, e = 86°, f = 37°$

Solve for the unknown angles. (DOK 2)

1.

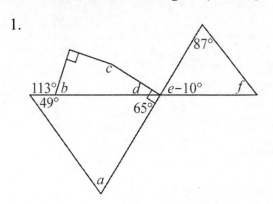

2.

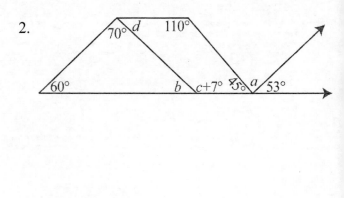

Chapter 10 Review

Find the measures. (DOK 1)

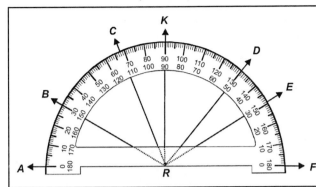

1. What is the measure of ∠DRA?

2. What is the measure of ∠CRF?

3. What is the measure of ∠ARB?

Use the following diagram for questions 4–14. (DOK 2)

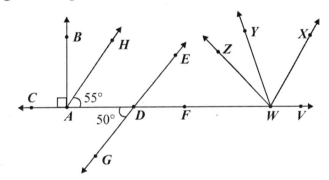

4. Which angles are supplementary angles to ∠EDF?

5. What is the measure of ∠GDF?

6. Which two angles are right angles?

7. What is the measure of ∠EDF?

8. Which angle is adjacent to ∠BAD?

9. Which angle is a complementary angle to ∠HAD?

10. What is the measure of ∠HAB?

11. What is the measure of ∠CAD?

12. What kind of angle is ∠FDA?

13. What kind of angle is ∠GDA?

14. Which angles are adjacent to ∠EDA?

Solve for the unknown angles in the figure below. (DOK 2)

15.

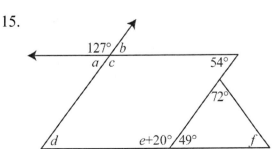

Chapter 10 Test

1 What type of angle is shown below?

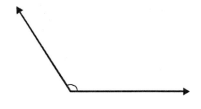

A right
B acute
C obtuse
D straight

(DOK 1)

Use the figure below to answer the following two questions.

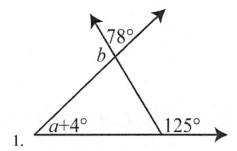

1.

2 What is the measure of angle a?

A 43°
B 55°
C 78°
D 102°

(DOK 2)

3 What is the measure of angle b?

A 47°
B 55°
C 78°
D 102°

(DOK 2)

4 In the diagram below, which two angles are adjacent?

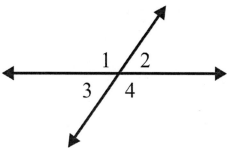

A ∠1 and ∠2
B ∠2 and ∠3
C ∠1 and ∠4
D Both A and B are correct.

(DOK 2)

5 What is the sum of two complementary angles?

A 180°
B 45°
C 90°
D 360°

(DOK 1)

6 What is the measure of an angle that is supplementary to 87°?

A −42°
B 3°
C 273°
D 93°

(DOK 2)

Chapter 11
Plane Geometry

This chapter covers the following CC Grade 7 standards:

	Content Standards
Geometry	7.G.1, 7.G.2, 7.G.6

11.1 Types of Triangles (DOK 1)

right triangle
contains 1 right $\angle$

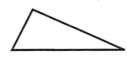

acute triangle
all angles are acute
(less than 90°)

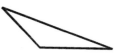

obtuse triangle
one angle is obtuse
(greater than 90°)

equilateral triangle
all three sides equal
all angles are 60°

scalene triangle
no sides equal
no angles equal

isosceles triangle
two sides equal
two angles equal

Name the type of triangles described below. (DOK 1)

1. What kind of triangle has only two sides measuring the same length?

2. What kind of triangle has three angles measuring 60°?

3. What kind of triangle has angles measuring 37°, 85°, and 58°?

4. This triangle has one angle measuring 90°.

5. This triangle has one angle measuring 120°.

6. This triangle has three angles less than 90°.

7. What kind of triangle has side measures of 3 inches, 4 inches, and 6 inches?

8. What kind of triangle has angles measuring 90°, 60°, and 30°?

11.2   Types of Quadrilaterals (DOK 1)

A **quadrilateral** is any 4 sided figure. (quad means 4 and lateral means side.)

Types of Quadrilaterals

Square
4 equal sides
Four 90° angles

Rectangle
2 sets of equal sides
Four 90° angles

Rhombus
4 equal sides
2 sets of equal angles

Trapezoid
1 pair of
parallel sides

Irregular
Any side measures
Any angle measures

Name the quadrilateral in each description below. (DOK 1)

1. What quadrilateral has four 90° angles, but not all sides are congruent?

2. Which quadrilateral has four equal angles and sides?

3. Which quadrilateral might have sides measuring 3", 5", 8", and 6"?

4. Which quadrilateral has four sides that are congruent, but not all angles are congruent?

5. Which quadrilateral has one set of parallel sides and two sides that are not parallel?

6. Which quadrilateral has two sides measuring 3 inches and two sides measuring 8 inches, with four 90° angles?

11.3   Quadrilaterals and Their Properties (DOK 1)

A **quadrilateral** is a polygon with four sides. A **parallelogram** is a quadrilateral in which both pairs of opposite sides are parallel. The following properties of parallelograms are given without proof:

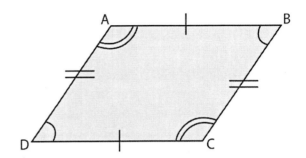

1. **Both pairs of opposite sides are parallel.**

2. **The opposite sides are congruent.**

3. **The opposite angles are congruent.**

4. **Consecutive angles are supplementary.**

A **rectangle** is a parallelogram with four right angles. It follows that a rectangle has all of the properties listed above, plus all four angles are 90°. A **rhombus** is a parallelogram with four congruent sides. A rhombus has all the properties of a parallelogram, but both pairs of opposite sides are congruent as well as parallel. A **square** is a rhombus with four right angles. Therefore, a square has four congruent sides and four congruent angles (each 90°). As you can see, a square is also a quadrilateral, a parallelogram, a rectangle, and a rhombus. These properties plus the four right angles make it a square.

A **trapezoid** is a quadrilateral with only one pair of parallel sides. The parallel sides are called bases, and the other two sides are called legs. The legs do not have to be the same length.

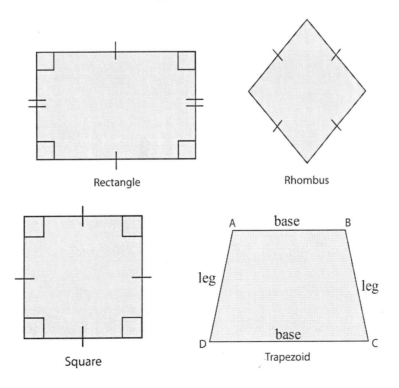

Rectangle

Rhombus

Square

Trapezoid

11.4   Types of Polygons (DOK 1)

Polygon means *many sided.* A *regular* polygon has sides that are all equal in measure and angles that are congruent. A polygon may have sides that are not equal in size.

Triangle - 3 sides and 3 angles.

Square - all 4 sides are equal, 2 pairs of sides parallel to each other. All 4 angles equal and are right angles.

Rectangle - 2 pairs of equal and parallel sides, all 4 angles are right angles. One pair of sides may be longer than the other pair of sides.

Trapezoid - 2 sides are made up of 1 pair of parallel lines that are opposite each other, and 2 sides that are not parallel to each other.

Rhombus - All 4 sides are equal. 2 pairs of sides that are parallel to each other. 2 pairs of equal angles.

Parallelogram - 2 pairs of parallel lines that are opposite each other, making up two pairs of equal sides and 2 pairs of equal angles.

Pentagon - 5 sides. A regular pentagon has 5 equal sides and 5 equal angles.

Hexagon - 6 sides. A regular hexagon has 6 equal sides and 6 equal angles, 3 pairs of sides are parallel and opposite each other.

Heptagon - 7 sides. A regular heptagon has 7 equal sides and 7 equal angles.

Octagon - 8 sides. A regular octagon has 8 equal sides and 8 equal angles, 4 pairs of sides are parallel and opposite each other.

Nonagon - 9 sides. A regular nonagon has 9 equal sides and 9 equal angles.

Decagon - 10 sides. A regular decagon has 10 equal sides and 10 equal angles, 5 pairs of sides are parallel and opposite each other.

Fill in the names of each polygon below. Refer to the descriptions above to help you. (DOK 1)

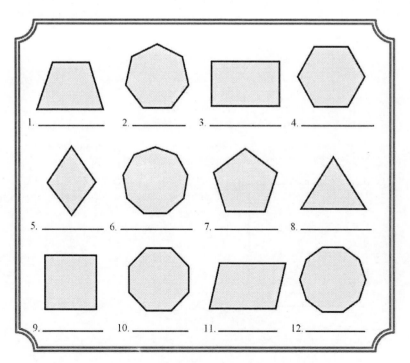

11.5   Drawing Shapes (DOK 2)

Example 1:   Draw an isosceles triangle freehand.

Step 1:   Recall the properties of an isosceles triangle.
In an isosceles triangle, two sides and two angles are equal, or congruent.
Draw a base, $\overline{MN}$. Draw a horizontal straight line. Label both ends of the line.

Step 3:   Mark a point above the middle of base $\overline{MN}$.
This point marks where the legs should meet in order for them to have congruent lengths and have congruent angles from the base.

Step 4:   Draw two legs, $\overline{MO}$ and $\overline{NO}$.
Connect both ends of the base to the point O, drawing both legs as congruent as possible.

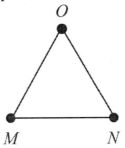

Step 5:   Draw hash marks and arcs to show congruence.
Drawing hash marks on the legs show they are congruent in length and drawing arcs on $\angle NMO$ and $\angle MNO$ show those angles are congruent.

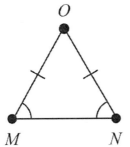

Example 2:   Draw a rectangle using the properties of rectangles freehand.

Step 1:   Recall the properties of a rectangle.
In a rectangle, there are two pairs of equal and parallel sides and all four angles are right angles.

Step 2:   Draw a side, $\overline{AB}$. Draw a horizontal straight line. Label both ends of the line.

Step 3:   Draw a side, $\overline{BC}$
Draw a line perpendicular to the line segment $\overline{AB}$ that starts at point B.
Label the end of the new line segment.
Draw a small square in $\angle ABC$ indicating that angle measures $90°$.

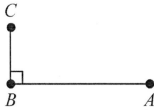

Step 4:   Draw a side, $\overline{CD}$
Draw a line perpendicular to $\overline{BC}$ and parallel to $\overline{AB}$. Draw line $\overline{CD}$ as congruent to $\overline{AB}$ as possible.
Label the end of the new line segment.
Draw a small square in $\angle DCB$ indicating that angle measures $90°$.

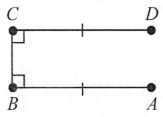

Step 5:   Draw a side, $\overline{DA}$.
Connect points D and A with a straight line, drawing line $\overline{DA}$ as congruent to $\overline{BC}$ as possible.
Draw small squares in $\angle CDA$ and $\angle DAB$ indicating those angles measure $90°$.
Draw one hash mark on $\overline{AB}$ and $\overline{CD}$ and two hash marks on $\overline{BC}$ and $\overline{DA}$ to indicate they are congruent.

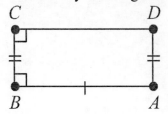

Example 3:   Draw an acute triangle freehand.

Step 1:   Recall the properties of an acute triangle.

An acute triangle has three acute angles meaning they are less than 90°.

Step 2:   Draw a base, $\overline{EF}$.

Draw a horizontal straight line.

Label both ends of the line.

Step 3:   Draw a leg, $\overline{EG}$.

Draw leg $\overline{EG}$ so that the angle it forms with base $\overline{EF}$ is less than 90°.

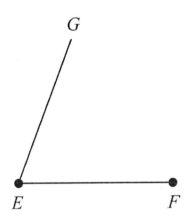

Step 3:   Draw a leg, $\overline{GF}$.

Connect point G and F to draw the other leg, $\overline{GF}$, making sure $\angle EGF$ and $\angle GFE$ are less than 90°.

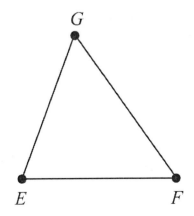

Example 4: Draw a triangle with angles that measure 45°, 45°, and 90° with one side measuring 3 cm using a protractor and ruler.

Step 1: Use a ruler to draw a line 3 cm long.
Label both ends of the line segment.

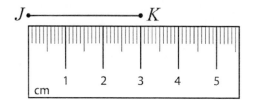

Step 2: Use a protractor to measure a 45° angle from one side of base $\overline{JK}$.
Line up the central point of the protractor with point J. At the 45° angle line, draw a mark.

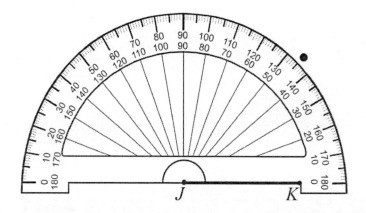

Step 3: Use the same process as in step 2 to construct a line from point K at a 45° angle.

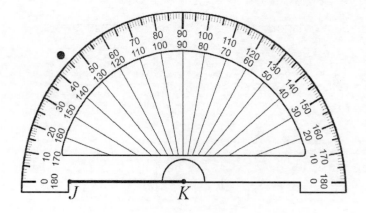

Step 4:   Move the protractor and draw a straight line with the ruler starting at point J and extending through the mark.
Do the same process starting at point K.
Label the point of intersection.

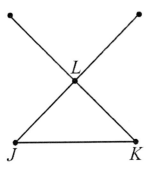

Erase the lines that extend past point L and draw hash marks and arcs to show congruency.

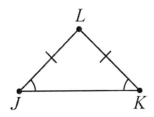

Measure $\angle JLK$ with the protractor to check that it is $90°$

Note: If $\angle JLK$ does not measure $90°$, re-measure the other two angles to check that they are $45°$.

Example 5:   Draw a regular pentagon with all 5 sides measuring 3 cm using a protractor and ruler.

Step 1:   Recall the properties of a regular pentagon.

A regular pentagon has 5 equal sides and 5 equal angles.

Step 2:   Draw a side $\overline{ST} = 3$ cm

Use a ruler to draw a line 3 cm long.

Label both ends of the line segment.

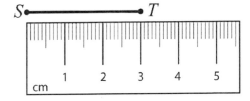

Step 3:   Use a protractor to measure a 108° angle from one side of base $\overline{ST}$.

Line up the central point of the protractor with point S. At the 108° angle line, draw a mark.

Note: All interior angles of a regular pentagon measure 108°. The formula for interior angles of regular polygons will be taught later.

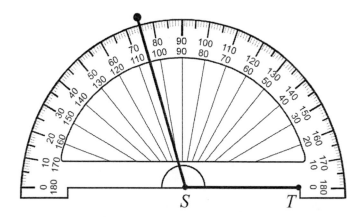

Step 4:   Move the protractor and use the ruler to measure 3 cm of the new line.

Erase the part of the line which extends past 3 cm and label the end of the new line.

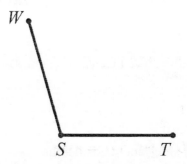

Repeat steps 3 and 4 until there are 5 congruent sides and angles.

Draw hash marks and arcs to show congruency.

If the last side drawn does not connect with point T, re-measure the sides and angles to verify they are all congruent.

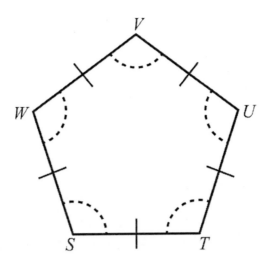

Read the following descriptions and draw the shape freehand (with no ruler or protractor). Note: Use hash marks, and arcs to show congruence between sides and angles and small squares to show right angles. Draw each figure with the given information. (DOK 2)

1. Obtuse Triangle, $\triangle ABC$

2. Scalene Triangle, $\triangle DEF$

3. Right Triangle, $\triangle GHI$

4. Square, $JKLM$

5. Regular Hexagon, $NOPQRS$

Read the following descriptions and draw the shape using a protractor and ruler.

6. Triangle, $\triangle TUV : m\angle VTU = 90°, m\angle TVU = 60°, \overline{TU} = 2$ cm

7. Regular Octagon $ABCDEFGH$: All angles measure $135°$, each side measures 2 cm

8. Triangle, $\triangle XYZ$: All angles measure $60°$, all sides measure 3 cm

9. Parallelogram $JKLM$: Two sides measuring 2 cm and two sides measuring 3 cm

10. Isosceles Triangle, $\triangle NOP$: Two angles measuring $30°$

11.6   Area of Squares and Rectangles (DOK 1)

Area is always expressed in square units, such as sq in (in²), sq m (m²), or sq ft (ft²).

The area, (A), of squares and rectangles equals length (l) times width (w). $A = l \times w$.

Example 6:
4 cm

4 cm

$A = lw$
$A = 4 \times 4$
$A = 16 \text{ cm}^2$

If a square has an area of 16 cm², it means that it will take 16 squares that are 1 cm on each side to cover the area that is 4 cm on each side.

Find the area of the following squares and rectangles using the formula $A_{\text{rectangle}} = lw$. (DOK 1)

1.   10 ft

10 ft

2.   5 cm

2 cm

3.   4 in

9 in

4.   9 in

20 in

5.   6 ft

6 ft

6.   10 cm

5 cm

7.   4 ft

2 ft

8.   5 in

8 in

9.   12 ft

12 ft

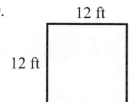

10.   7 cm

12 cm

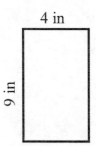

11.   1 ft

8 ft

12.   6 cm

7 cm

11.7   Area of Triangles (DOK 2)

Example 7:   Find the area of the following triangle.
The formula for the area of a triangle is as follows:

$$A = \frac{1}{2} \times b \times h$$

A = area
b = base
h = height or altitude

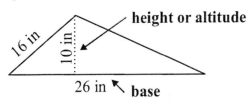

height or altitude

base

Step 1:   Insert the measurements from the triangle into the formula: $A = \frac{1}{2} \times 26 \times 10$

Step 2:   Cancel and multiply.
$$A = \frac{1}{\overset{}{\underset{1}{2}}} \times \frac{\overset{13}{\cancel{26}}}{1} \times \frac{10}{1} = 130 \text{ in}^2$$

Note: **Area is always expressed in square units such as sq in, sq ft, or sq m.**

Find the area of the following triangles. Remember to include units. (DOK 2)

1.

5 in
3 in
4 in

5.

3 ft→
2 ft

9.

2 ft
2 ft

2.

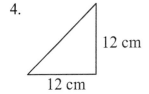

7 cm
6 cm
height
12 cm

6.

20 cm
16 cm

10.

5 ft
4 ft
6 ft

3.

6 ft
9 ft
7 ft

7.

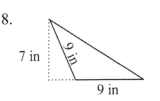

8 m
height→ 7 m
15 m

11.

10 ft
12 ft
15 ft

4.

12 cm
12 cm

8.

7 in
9 in
9 in

12.

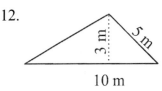

3 m
5 m
10 m

11.8   Area of Trapezoids and Parallelograms (DOK 2)

Example 8:   Find the area of the following parallelogram.

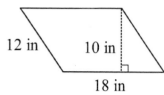

The formula for the area of a parallelogram is $A = bh$.
A = area
b = base
h = height

Step 1:   Insert measurements from the parallelogram into the formula: $A = 18 \times 10$.

Step 2:   Multiply. $18 \times 10 = 180$ in^2

Example 9:   Find the area of the following trapezoid.
The formula for the area of a trapezoid is $A = \frac{1}{2}h(b_1 + b_2)$. A trapezoid has two bases that are parallel to each other. When you add the length of the two bases together and then multiply by $\frac{1}{2}$, you find their average length.

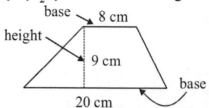

A = area
b = base
h = height

Insert the measurements from the trapezoid into the formula and solve:
$\frac{1}{2} \times 9\,(8 + 20) = 126$ cm^2

Find the area of the following parallelograms and trapezoids. (DOK 2)

1.
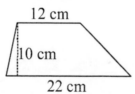

4.

7.

2.

5.

8.

3.
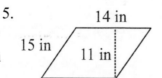

6.
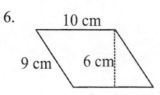

9.

11.9   Area of Composite Figures (DOK 2)

A **composite figure** is made up of more than one polygon.

Example 10:   Find the area of the composite figure below.

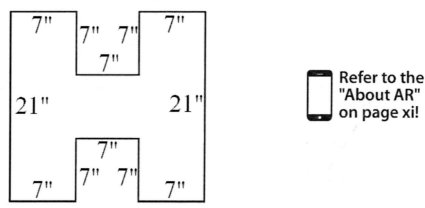

Refer to the
"About AR"
on page xi!

Step 1:   Determine how the figure could be separated into parts and find the area of each part.

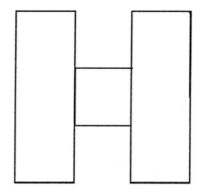

The figure can be separated into 2 rectangles measuring 21 inches by 7 inches each and one 7 inch square.

$21 \times 7 = 147$ sq in and $7 \times 7 = 49$ sq in.

Step 2:   Add up the area of each part:

One rectangle	147	sq in
One rectangle	147	sq in
One square	49	sq in
Total	343	sq in

The total area is 343 sq in.

Find the area of each of the composite figures below. (DOK 2)

1.

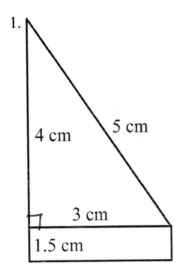

3.

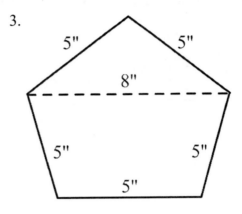

2.

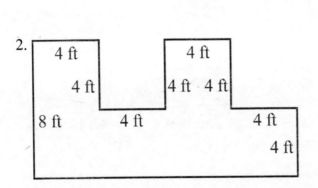

4.

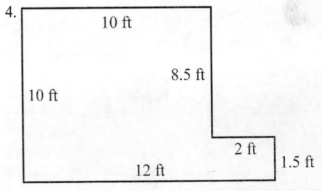

11.10   Area Word Problems (DOK 2)

Solve the following area word problems. (DOK 2)

1. Stephen is building a fence around his family's garden. The fence will be 5 ft wide and 7 ft long. What area does the garden occupy?

2. Jenna has a pizza in the shape of a regular hexagon with sides of 4 in. She wants to share it with five of her friends, so she cuts the pizza into 6 triangular slices. If the distance from an edge of the pizza to the center is 6 inches, what is the area of one slice of pizza?

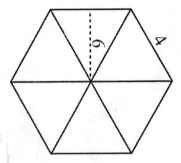

3. Using the pizza as described in the previous problem, what is the area of the entire hexagonal pizza?

4. What is the area of a pentagon composed of a square with side length 5 and a triangle with base 5 and height 3?

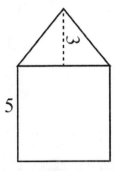

5. A small square lies inside a larger square. The large square has a side length of 9 and the small square has a side length of 4. What is the area of the region outside of the small square but inside the large square?

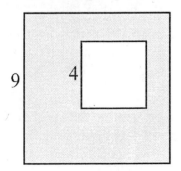

11.11   Scale Drawings (DOK 2)

You can use scale drawings to figure the size of an object. For instance, if you draw a grid with one inch squares on a clear plastic sheet, you can tape it to a window and draw the landscape you see square by square, by drawing the same number of squares on your paper. Perhaps a 20 foot tree will take up 3 squares. It's a great way to get your scale and proportion correct on your drawing. You can use that same piece of gridded plastic to draw from a picture by laying the grid over a photo and copying each square onto your drawing paper that you have drawn gridlines like on the graphic below.

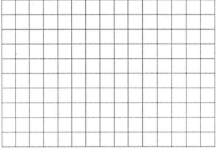

A scale drawing will have a table on it with equivalent measurements. For instance: 1 inch = 1 foot. You know there are twelve inches to a foot, so the scale is a $\frac{1}{12}$ ratio.

Example 11:   Suppose that on the grid above, there is a figure that measures 4 inches. How tall is that figure in the real world?

Step 1:   Set up a proportion.
Scale: $\dfrac{1 \text{ inch}}{1 \text{ foot}} = \dfrac{1}{12}$ The figure: $\dfrac{4}{x}$

Step 2: $\dfrac{1}{12} = \dfrac{4}{x} \Rightarrow x = 12 \times 4 = 48$

$x = 48$ The 4 inch drawing represents a 48 inch figure.

Set up proportions for each of the following problems and solve. (DOK 2)

1. Cherie is drawing her backyard using a gridded plastic sheet. Her grid on the sheet is in 1 inch squares. It takes 6 squares high by 3 squares wide to draw the backyard shed. She knows the shed is 6 feet tall. How much does each of her 1 inch grids represent?

2. On a scale drawing of a new toy being made by the Lots-a-Fun Toy Company, 1 grid unit = 2 inches. The drawing of the wheel takes up 2 grids tall and 2 grids wide. How tall is the wheel?

3. Merilee is painting a mural of birds and flowers on her bedroom wall. The drawing she is using for her inspiration shows a bird 3 inches tall. Merilee wants to paint the bird 6 feet tall on her wall. If Merilee's gridlines on the picture of the bird measure 1 inch square, how much does each square represent on her mural?

4. Rob is taking a mechanical drawing class. The teacher has Rob drawing an auto part to scale. Each quarter inch represents 1 inch on the auto part. If the auto part is 6 inches long, how long will Rob make the auto part on his drawing?

11.12   Solving Word Problems Using Scale Drawings (DOK 2)

Carefully read each problem below and solve. (DOK 2)

1. On a scale drawing of a quilt, one quilt square is 9 by 9 inches. This represents 1 square inch on the scale drawing. If the final length of the quilt is 108 inches, how many inches is the length of the quilt on the scale drawing?

2. On a map of a shopping mall, the Isabella Ice Cream stand is 12 cm from the point on the map to the Hotsie Hot Dog stand. The scale on the map says 2 cm = 30 feet. How far apart are the two stands, assuming there is a direct route?

3. The scale on a map of the ground floor of a dinosaur museum is $\frac{1}{3}$ inches equals 3 feet. It is 7 inches on the map from the gift shop to the T-Rex. How many feet does that translate to?

4. The scale on a map of the new sports arena is 1 inch equals 100 feet. The length of the football field in the arena is 3 inches in length on the map. How long is the football field?

5. On a map of North Carolina, 27 miles equals one centimeter. The distance from Hickory, North Carolina to Lumberton, North Carolina is 189 miles. What is the distance in centimeters between the cities on the map?

6. Hank and his horse, Cowpoke, are riding out to pan for gold. The map Hank has says 1 inch equals 10 miles. Using the top of his index finger as about one inch, Hank realizes they need to ride 45 miles south and 12.5 miles west to get to their location. How many inches is that distance represented on Hank's map? Give your answer in number of miles and direction.

7. Rodney and Rodina are traveling to their mother's house. According to the map, they need to travel 3.5 inches north and then 2 inches straight west. One inch is equal to 20 miles. Find the number of miles they will travel.

8. Mr. Tolafson drew a scale drawing of his model railroad world to use in case his grandkids and dogs moved things out of place. The scale on his drawing says $\frac{1}{4}$ inch equals 3 inches. If the water tower is $\frac{3}{4}$ inch from the bridge that goes over a ravine, how far is the water tower from the bridge in reality? If the herd of plastic buffalo is 1.5 inches from the top of the mountain on the drawing, how far apart are they in reality? If the train station is 24 inches from the Livery Stable in reality, how far apart are they on the scale drawing?

11.13   Solving Scale Drawings Using Figures (DOK 2)

Similar shapes are shapes that have the same angles and proportions, but have different sizes. Similar shapes are used in scale drawings. The factor by which a side length increases or decreases is called the **scale factor**. If the scale for a drawing is $\frac{1}{2} = 2$, then the scale factor is 4 (because you have to multiply $\frac{1}{2}$ by 4 to equal 2). The area increases by a factor of 42, or $4 \times 4 = 16$. The area always increases by a factor of the scale factor multiplied by itself. If the drawing is of a circle, the radius would increase by a factor of 4 and the area would increase by a factor of 16.

Example 12:   The scale for a drawing of a field is 1 in = 3 ft. Draw the figure according to the scale and label the new diagram with the actual measurements. What will be the actual area of the field?

5 in

$2\frac{1}{2}$ in $2\frac{1}{2}$ in

5 in

Step 1:   The scale is 1 to 3, so you will need to draw each of sides 3 times bigger than what is shown.

Step 2:   Multiply each of the side lengths by 3 and label the new figure.

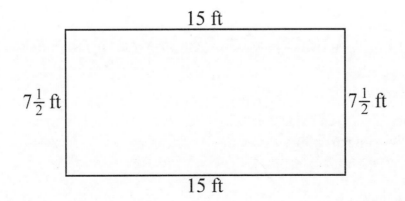

15 ft

$7\frac{1}{2}$ ft $7\frac{1}{2}$ ft

15 ft

Step 3:   Find the area of the field. $15 \times 7.5 = 112.5$ sq. ft. (Note that this is 9 times the $7\frac{1}{2}$ ft area of the smaller figure.)

Solve the following problems. (DOK 2)

1. The scale for the drawing below is $\frac{1}{2}$ cm = 2 cm. Draw the larger figure and label the sides. What is the area of the larger figure? By what factor did the area increase from the smaller figure?

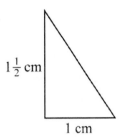

$1\frac{1}{2}$ cm

1 cm

2. The scale for the drawing below is 1 in = 2 in. Draw the larger figure and label the sides. By what factor did the sides increase? By what factor did the area increase?

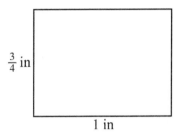

$\frac{3}{4}$ in

1 in

3. Jenny's father made her a doll house which was modeled after the blueprint of her grandmother's home. The blueprint is 24 inches by 45 inches. The scale factor is 1.25. What are the dimensions of the doll house?

4. A photograph was enlarged and framed. The framed photo is 12 inches by 20 inches. The scale factor is 4. What are the original dimensions of the photograph?

5. If the scale for a drawing of a tower is $\frac{1}{3}$ in = 24 in, by what factor will the height of the actual tower increase?

6. The scale for a drawing of a circular window is 1 in = 8 in. By what factor will the area of the drawing increase when the window is actually constructed?

11.14   Going Deeper into Plane Geometry (DOK 3)

Use the drawing below of a floor plan to answer questions 1–4. Show your work for each step. (DOK 3)

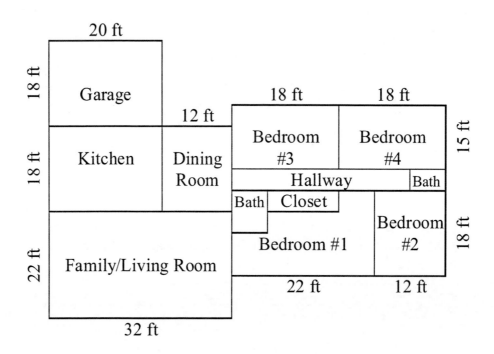

Kitchen, dining room, and family/living room

1. Find the total area of the kitchen, dining room, and the family/living room.

2. If the family that lives in this house decides to remodel and adds 6 feet to the width of the family/living room, how many additional square feet will they add on to the room?

Bedrooms

3. Find the total area of all the bedrooms. Include the bath and closet as part of bedroom #1.

4. If the family decided to take away the wall between bedrooms 3 and 4, what will the area of the new, larger bedroom be?

5. The drawing to the right is a scale drawing of a patio area that will be added to the yard of a house. The scale is 1 inch = 4 feet. Find the area of the patio in square feet. Show your work.

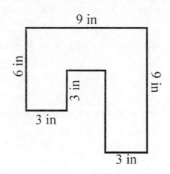

Chapter 11 Review

Answer the questions about shapes below. (DOK 1)

1. What kind of triangle has only two sides and two angles equal?

2. What kind of quadrilateral has 2 sets of equal sides and 4 equal angles?

3. What kind of polygon has 5 sides?

4. What kind of polygon has 7 sides?

Solve the area problems below. (DOK 2)

5. What is the area of a rectangle measuring 8 inches by 12 inches?

6. What is the area of a square which measures 8 inches on each side?

7. Find the area of the following parallelogram. $A_{\text{parallelogram}} = bh$

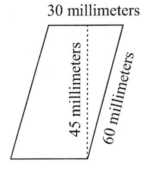

30 millimeters

8. What is the area of the parallelogram below? $A = bh$

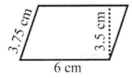

9. Find the area of the composite figure below.

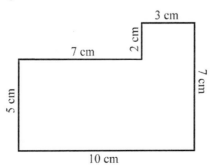

Use what you know about ratios to solve. (DOK 2)

10. A print of a famous painting is $\frac{1}{3}$ the size of the original. If the original is 9 feet long, how long is the print?

11. A drawing of a whale's skeleton shows a scale of 1 inch : 8 feet. How long is the whale if the skeleton is drawn 8.5 inches long?

12. A map's scale shows 2.5 cm = 1 km. How far are two points on the map that are 2.3 km in real life?

13. Elisa is drawing a picture of her German Shepherd to a scale of 1 foot : 6 inches. If her dog is $2\frac{1}{2}$ feet tall, how tall will be Elisa's drawing?

14. Jake has a model airplane to scale of $\frac{1}{4}$ inch : 12 inches. If the propeller is 30 inches long on the airplane, how long is the propeller on the scale drawing?

15. Alex is using a map with a scale of 1 inch = 2 miles. The distance on the map from Alex's house to the planetarium is $7\frac{1}{2}$ inches. How far is the planetarium from Alex's house?

Solve the following scale drawing problems. (DOK 2)

16. Tina wants to enlarge a photograph of her family to put in a frame. If the scale is 1 inch = $2\frac{1}{2}$ inches and the dimensions of the photo are 2 in by $2\frac{1}{2}$ in, what will the dimensions of the final photo be?

17. The scale for the drawing below is 1 cm = 3 cm. Draw the larger figure.

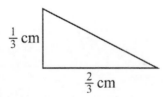

$\frac{1}{3}$ cm

$\frac{2}{3}$ cm

The drawing below and to the right is the dimensions of a family living room. Show your work for each step while answering the question about remodeling the living room (DOK 3)

18. The scale drawing to the right shows the dimensions of the Dugan's living room. The scale is 1 inch = 4 feet. Mr. Dugan drew an additional 2 inches to both dimensions of the scale drawing, to increase the size of the living room. How many additional square feet will be added to the living room?

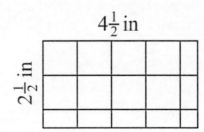

$4\frac{1}{2}$ in

$2\frac{1}{2}$ in

Chapter 11 Test

1 What kind of triangle has all sides and angles the same size?

A right
B acute
C obtuse
D equilateral

(DOK 1)

2 Which quadrilateral is the figure below?

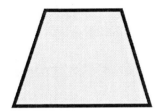

A square
B trapezoid
C rectangle
D rhombus

(DOK 1)

3 What polygon has nine sides?

A triangle
B pentagon
C octagon
D nonagon

(DOK 1)

4 What polygon has 2 pairs of parallel lines that are opposite each other, making up 2 pairs of equal sides and 2 pairs of equal angles?

A heptagon
B pentagon
C parallelogram
D decagon

(DOK 1)

5 Find the area of the rectangle below. $A = lw$

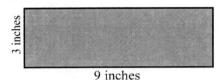

A 12 sq in
B 24 sq in
C 25 sq in
D 27 sq in

(DOK 2)

6 What is the area of the triangle below? $A = \frac{1}{2}bh$

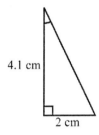

A 6.1 sq cm
B 4.1 sq cm
C 8.2 sq cm
D 2.05 sq cm

(DOK 2)

7 What is the area of the parallelogram below? $A = bh$

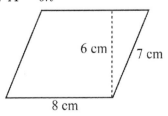

A 14 sq cm
B 56 sq cm
C 48 sq cm
D 24 sq cm

(DOK 2)

8 What is the area of the trapezoid below?
$A = \frac{1}{2}h(b_1 + b_2)$

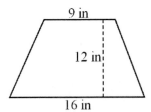

A 37 sq in
B 75 sq in
C 96 sq in
D 150 sq in

(DOK 2)

9 What is the area of the composite figure below? $A = lw$

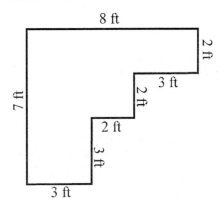

A 30 sq ft
B 35 sq ft
C 40 sq ft
D 56 sq ft

(DOK 2)

10 Sharisse uses a map with a scale of 1 inch : 2 miles. She finds the library is $3\frac{1}{4}$ inches from her house. How far will Sharisse have to travel from her house to get to the library?

A 6.5 miles
B 3.25 miles
C 3.25 inches
D 6.5 inches

(DOK 2)

11 Two cities are 3.5 inches apart on a map. The map scale shows 0.5 in = 10 mi. How far apart are the cities in real life?

A 10 miles
B 350 miles
C 35 miles
D 70 miles

(DOK 2)

12 On the architectural plans for a school gymnasium, $\frac{1}{4}$ inch represents 2 feet. The length of the gymnasium on the drawing is currently 16 inches. The remodel on the gymnasium will lengthen the gym by 32 feet. How many inches long will the drawing be for the remodeled gymnasium?

A 80 inches
B 40 inches
C 20 inches
D 10 inches

(DOK 3)

13 A volcano in Russia is nearly 4800 m high. Daniel is building a model of the volcano for a science project. He uses the scale 1 cm = 120 m. How tall is Daniel's model?

A 40 cm
B 40 m
C 50 cm
D 30 cm

(DOK 2)

14 A map shows a scale of 1 cm : 5 km. What would the map distance be if the actual distance is 29 km?

A 4.8 cm
B 5 cm
C 5.8 cm
D 6 cm

(DOK 2)

Chapter 12
Circles

This chapter covers the following CC Grade 7 standard:

	Content Standard
Geometry	7.G.4

12.1   Parts of a Circle (DOK 1)

A circle can have a **radius**, **diameter**, **chord**, **center**, and **circumference**. A proportional relationship exists between the diameter and circumference in a circle and the unit rate is π.

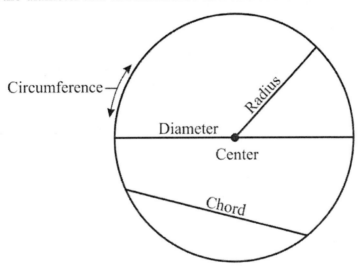

Radius - A line segment from the center of a circle to the edge of a circle. A radius is one half the length of a diameter.

Diameter - A line segment passing through the center of a circle from one side of the circle to the other. A diameter is twice as long as a radius.

Chord - A chord is a line segment that starts on the circumference and ends on the circumference. It does not need to go through the center of a circle.

Center - A center is at the exact midpoint of a circle in all directions.

Circumference - The distance around the outside of a circle.

12.2 Circumference (DOK 1)

The formula for the **circumference** of a circle is $C = 2\pi r$ or $C = \pi d$.
(C = circumference, r = radius, d = diameter)
The formulas are equal because the diameter is equal to twice the length of the radius.

Example 1:   Find the circumference of the circle below.

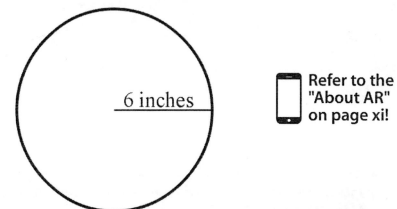

6 inches

Refer to the "About AR" on page xi!

Step 1:   Use the formula $C = 2\pi r$. Find the value of r. The radius is 6 inches.

Step 2:   Plug the values into the formula and solve. Do not multiply π, just multiply the whole numbers, 2 and 6. (This is called leaving the answer in terms of π.)
$C = 2 \times \pi \times 6$

Answer: $C = 12\pi$ inches

Example 2:   Find the circumference of the circle below.

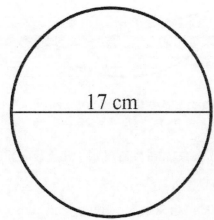

17 cm

Step 1:   Use the formula $C = \pi d$. Find the value of d. The diameter is 17 centimeters.

Step 2:   Plug the values into the formula and solve.
$C = \pi \times 17$

Answer: $C = 17\pi$ cm

Find the circumference of the following circles. Leave all answers in terms of π. (DOK 1)

1.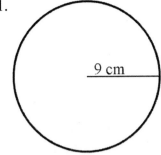
9 cm

2.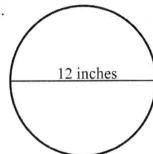
12 inches

3.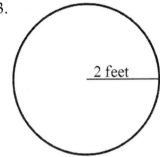
2 feet

4.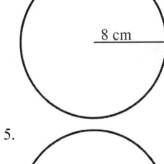
8 cm

5.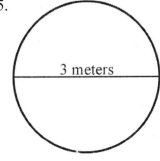
3 meters

6.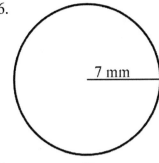
7 mm

7.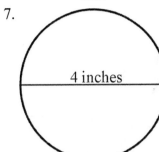
4 inches

8.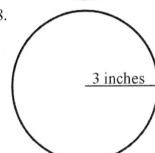
3 inches

9.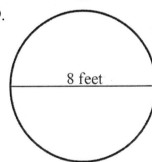
8 feet

10.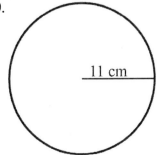
11 cm

12.3   Area of a Circle (DOK 1)

The **area** of a circle is the number of square units of measure that will fit inside a circle. The formula for the **area of a circle** is $A = \pi r^2$.

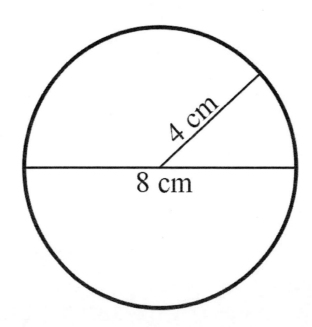

Example 3:   Find the area of the circle above.

Step 1:   Use the formula $A = \pi r^2$. Find the value of r.

Step 2:   Plug the values into the formula and solve. $A = \pi \times 4^2$.

Answer: 16π cm^2

Example 4:   Find the area of a circle with a radius of 8 inches.

Step 1:   Use the formula $A = \pi r^2$. Find the value of r.

Step 2:   Plug the values into the formula and solve. $A = \pi \times 8^2$.

Answer: 64π cm^2

Find the area of the following circles. Leave all answers in terms of π. (DOK 1)

1.

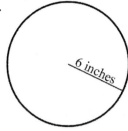

$A =$ _____

2.

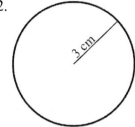

$A =$ _____

3.

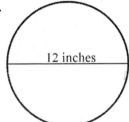

$A =$ _____

4.

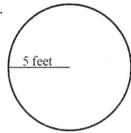

$A =$ _____

5.

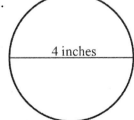

$A =$ _____

12.4   Relating Circumference and Area (DOK 2)

The circumference and area of a circle are closely related. To show how they are related, a ratio between circumference and area will be found.

To find circumference to area, divide the area formula by the circumference formula, $\dfrac{\pi r^2}{2\pi r}$.

π and r cancel out, leaving $\dfrac{A}{C} = \dfrac{r}{2}$

This means that if you multiply the circumference by $\dfrac{r}{2}$, you will have the area of the circle. Likewise, if you divide the area by $\dfrac{r}{2}$, you will have the circumference of the circle.

Solve the following circumference and area problems. (DOK2)

1. Suppose the area of a circle is 50 sq. cm and its radius is known to be 4 cm. What is the circumference of the circle?

2. Suppose the circumference of a circle is 44 cm and its radius is known to be 7 cm. What is the area of the circle?

12.5   Circle Word Problems (DOK 2)

Calculating the circumference or area of a circle can be useful in figuring out the amount of supplies needed to fence in a circular area, to find the amount of fabric needed to make a circular tablecloth, or to find the amount of flowers needed to cover a circular area, etc.

Example 5:   What is the area of a 4 foot diameter garden plot? $A = \pi r^2$

   Step 1:   Find the value of r. The radius is one half the diameter. The diameter is 4 feet.
$4 \div 2 = 2$.
The radius is 2 feet.

   Step 2:   Plug in the values of the formula and solve.
$A = \pi \times 2^2$.
$A = 4\pi$ square feet

   Answer: The area of a 4 foot diameter garden plot is 4π square feet.

Solve the following area and circumference problems. (DOK 2)

$$C = \pi d \text{ or } 2\pi r \qquad\qquad A = \pi r^2$$

1. The diameter of Diane's dining room table is 2 feet. What is the area of the table?

2. The radius of Ken's bicycle wheel is 10 inches. What is the circumference?

3. Teja needs to figure the amount of paint for a circle 14 feet in diameter. What is the area of the circle he is painting?

4. The kids are playing Snow Pie Tag on a pie with a radius of 12 feet. What is the circumference of the Snow Pie game?

5. Abby is making a round mouse maze for her pet mouse, The Amazing Zelda. The diameter of the maze is 24 inches. How large is the circumference of the mouse maze?

6. Mario is making a track for his model trains. He wants the track to go in a 4 foot diameter circle around the scenery he built for the train. What is the amount of track Mario will need rounded to the nearest whole number?

7. Lesia wants to replace only the top of a round footstool with new fabric. The diameter of the top of the footstool is 20 inches. How much fabric will Lesia need rounded to the nearest whole number?

12.6   Going Deeper into Circles (DOK 3)

Solve the problems below, showing your work for each step. Round your answers to the nearest tenth. Use $\pi = 3.14$. DOK 3)

1. Find the area of the <u>unshaded</u> portion of the circle below. The graphic in the middle is an exact quarter circle. The diameter of the full circle is 10 inches.

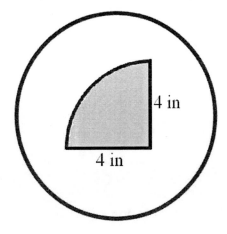

3. Find the area of the <u>shaded</u> portion of the circle below. Each of the wedges are exactly the same size. The diameter of the circle is 8 inches.

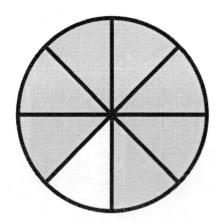

2. Find the area of the <u>unshaded</u> portion of the circle below. The radius is 6 cm. Each of the wedges are exactly the same size.

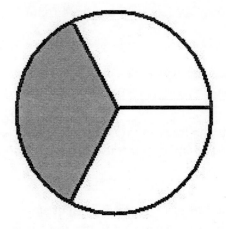

4. Find the area of the <u>shaded</u> portion of the circle below. The radius of the circle is 5 feet. Each of the wedges are exactly the same size.

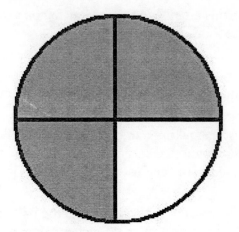

Chapter 12 Review

Use the diagram below to answer questions 1–4. (DOK 1)

1. What is the name of the part of the circle going from A to B?
2. What is the name of the part of the circle going from C to D?
3. What is the name of the part of the circle at point E?
4. What is the name of the part of the circle going from E to F?

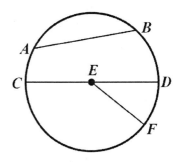

Fill in the chart below. Leave all answers in terms of π. (DOK 2)

$$C = \pi d \text{ or } 2\pi r \qquad A = \pi r^2$$

	Radius	Diameter	Circumference	Area
5.	3 inches	_____	_____	_____
6.	_____	8 cm	_____	_____
7.	_____	10 mm	_____	_____
8.	2 cm	_____	_____	_____
9.	1 foot	_____	_____	_____
10.	_____	2 yards	_____	_____
11.	6 cm	_____	_____	_____
12.	_____	14 inches	_____	_____

Follow the directions for the area of the problem below. Show your work for each step. (DOK 3)

13. Find the area of the <u>unshaded</u> portion of the circle below. The full circle has a diameter of 18 inches. Round your answer to the nearest tenth. Use $\pi = 3.14$.

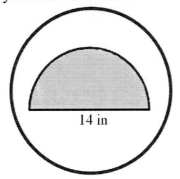

14 in

Chapter 12 Test

1 If the diameter of a circle is 18 inches, what is the radius?

A 36 inches
B 9 inches
C 18 inches
D 4.5 inches

(DOK 1)

2 If the radius of a circle is 24 feet, what is the diameter?

A 24 feet
B 36 feet
C 48 feet
D 12 feet

(DOK 1)

3 If the radius of a circle is 7 inches, what is the circumference of the circle? $C = 2\pi r$

A 7π inches
B 14π inches
C 21π inches
D 28π inches

(DOK 1)

4 What is the circumference of a circle with a diameter of 4 cm? $C = 2\pi r$

A 2π cm
B 8π cm
C 4π cm
D 16π cm

(DOK 1)

5 Barry has a table with a diameter of 60 inches. What is the radius of the table?

A 30 inches
B 120 inches
C 60 inches
D 376.8 inches

(DOK 1)

1. Use the circle below to answer questions 6–9.

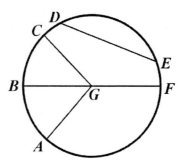

6 Which line segment represents a chord?

A line segment CG
B line segment GF
C line segment AG
D line segment DE

(DOK 1)

7 What is line segment BF called?

A center
B circumference
C diameter
D radius

(DOK 1)

8 Which line segment is defined the same as line segment CG?

A line segment AG
B line segment BF
C line segment DE
D line segment BC

(DOK 1)

9 If line segment AG is equal to 4 inches, how long is line segment BF?

A 4 inches
B 12 inches
C 8 inches
D 24 inches

(DOK 2)

1. **Use the circle below to answer questions 10–12.**

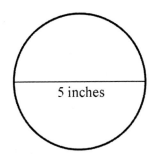

5 inches

10 What is the circumference of the circle?
$C = 2\pi r$

A 5π inches
B 2.5π inches
C 7π inches
D 10π inches

(DOK 1)

11 What is the radius of the circle?

A 15.7 inches
B 2.5 inches
C 5 inches
D 10 inches

(DOK 1)

12 What is the area of the circle? $A = \pi r^2$

A 2.5π inches^2
B 5π inches^2
C 25π inches^2
D 6.25π inches^2

(DOK 1)

13 What would you multiply the circumference of a circle by if you want to find the area of that circle?

A r

B $\dfrac{r}{4}$

C $\dfrac{r}{2}$

D $2r$

(DOK 1)

14 Find the area of the unshaded portion of the circle below. The large circle has a diameter of 12 inches. Use $\pi = 3.14$. Round your answer to the nearest tenth.

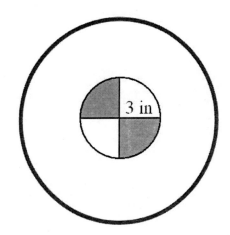

3 in

A 98.9 inches^2
B 423.9 inches^2
C 88.4 inches^2
D 84.4 inches^2

(DOK 3)

Chapter 13
Solid Geometry

This chapter covers the following CC Grade 7 standards:

	Content Standards
Geometry	7.G.3, 7.G.6

13.1   Solid Figures (DOK 1)

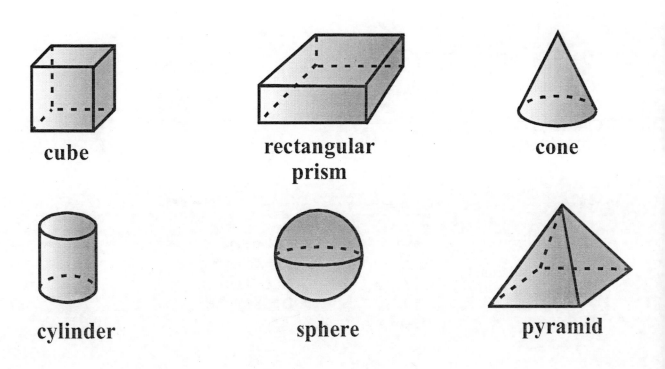

cube

rectangular
prism

cone

cylinder

sphere

pyramid

13.2   Cross Sections (DOK 1)

Cross Sections of Cubes

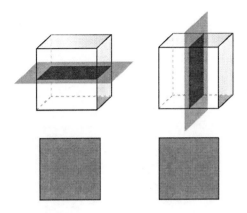

Cross Sections of Rectangular Prisms

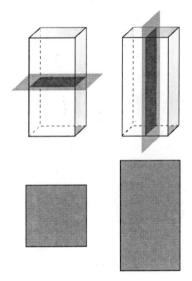

Cross Sections of Cones

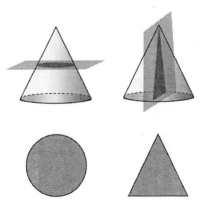

Cross Sections of Cylinders

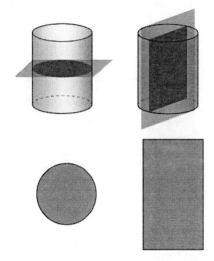

Cross Sections of Spheres

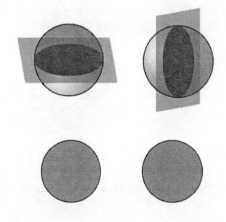

Cross Sections of Pyramids

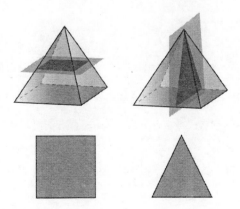

13.3   Understanding Volume (DOK 1)

Volume - Measurement of volume is expressed in cubic units such as in^3, ft^3, m^3, cm^3, or mm^3. The volume of a solid is the number of cubic units that can be contained in the solid.

First, let's look at rectangular solids.

Example 1:   How many 1 cubic centimeter cubes will it take to fill up the figure below?

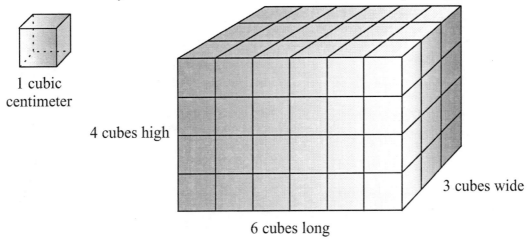

1 cubic
centimeter

4 cubes high

3 cubes wide

6 cubes long

To find the volume, you need to multiply the length times the width times the height.

Volume of a rectangular solid = length $\times$ width $\times$ height $(V = lwh)$.

$$V = 6 \times 3 \times 4 = 72 \text{ units}^3$$

13.4   Volume of Cubes and Rectangular Prisms (DOK 2)

A **cube** is a special kind of rectangular prism (box). Each side of a cube has the same measure. So, the formula for the volume of a cube is $V = s^3$ $(s \times s \times s)$.

Example 2:   Find the volume of the cube at right:

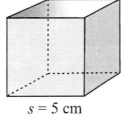

$s = 5$ cm

Step 1:   Insert measurements from the figure into the formula.

Step 2:   Multiply to solve. $5 \times 5 \times 5 = 125$ cm^3

Note: Volume is always expressed in cubic units such as in^3, ft^3, m^3, cm^3, or mm^3.

Answer each of the following questions about cubes. (DOK 2)

1. If a cube is 4 centimeters on each edge, what is the volume of the cube?

2. If the measure of the edge is doubled to 8 centimeters on each edge, what is the volume of the cube?

3. If the edge of a 4-centimeter cube is tripled to become 12 centimeters on each edge, what will the new volume be?

4. How many cubes with edges measuring 4 centimeters would you need to stack together to make a solid 16-centimeter cube?

5. What is the volume of a 3-centimeter cube?

6. Jerry built a 3-inch cube to hold his marble collection. He wants to build a cube with a volume 6 times larger. How much will each edge measure?

Find the volume of the following cubes. (DOK 2)

7.

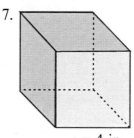

$s = 4$ in

8.

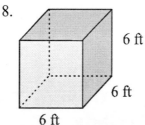

6 ft

6 ft

6 ft

9. 12 inches = 1 foot

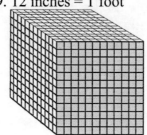

$s = 1$ foot

How many cubic inches are in a cubic foot?

You can calculate the volume (V) of a rectangular prism (box) by multiplying the length (l) by the width (w) by the height (h), as expressed in the formula $V = (lwh)$.

Example 3:   Find the volume of the box pictured here:

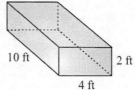

10 ft

2 ft

4 ft

Step 1:   Insert measurements from the figure into the formula.

Step 2:   Multiply to solve. $10 \times 4 \times 2 = 80$ ft^3

Note: **Volume is always expressed in cubic units such as** in^3**,** ft^3**,** m^3**,** cm^3**, or** mm^3**.**

Find the volume of the following rectangular prisms (boxes). (DOK 2)

1.
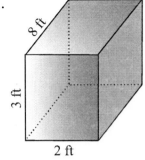

4.
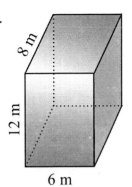

7.
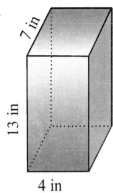

2.
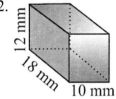

5.
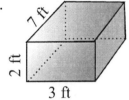

8.
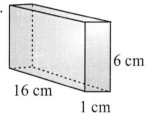

3.
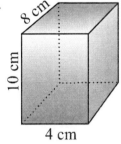

6.
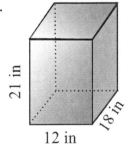

9.
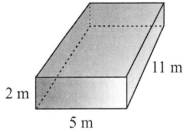

13.5   Volume of Other Prisms (DOK 2, 3)

Prisms can have any polygon as its face. To compute the volume of any prism, multiply the area of the face by the length of the prism.

Example 4:   A hexagonal prism has length 5 in. The area of the hexagon face is 24 in². What is the volume of the prism?

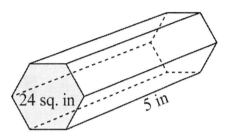

Multiply the area of the face by the length. $24 \times 5 = 120$

Answer: The volume of the prism is 120 in³.

Solve the following volume problems. (DOK 2, 3)

1. A hexagonal prism has a hexagon face with an area of 8.77 in². The length of the prism is 5 in long. What is the surface area of the hexagonal prism?

2. A heptagonal prism has a length of 12 cm. The area of the heptagon face is 37 cm². What is the surface area of the heptagonal prism?

3. A pentagonal prism has length 4.25 cm. The face of the prism is a pentagon with area 10 cm². What is the volume of the pentagonal prism?

4. A right triangular prism has a triangular face with sides 5, 12, and 13 in. The length of the prism is 3.5 in. What is the volume of this triangular prism?

5. A right triangular prism has length 6.5 cm. The face of the prism is a right triangle with legs 5.5 cm and 8 cm. What is the volume of the prism?

6. A decagonal prism has a length of 6 cm. Its decagon face has 10 equal sides with an area of 75 cm². What is the volume of the decagonal prism?

13.6   Volume of Pyramids (DOK 2)

To find the volume of a solid, insert the measurements given for the solid into the correct formula and solve. Remember, volumes are expressed in cubic units such as $\mathbf{in^3, ft^3, m^3, cm^3}$, or $\mathbf{mm^3}$.

Pyramids

$V = \dfrac{1}{3}Bh$ $B =$ area of rectangular base $\qquad$ $V = \dfrac{1}{3}Bh$ $B =$ area of triangular base

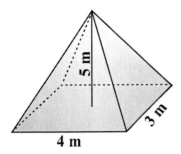

$V = \dfrac{1}{3}Bh$ $B = l \times w$ $\qquad\qquad\qquad$ $B = \dfrac{1}{2} \times 5 \times 4 = 10\,\text{ft}^2$

$V = \dfrac{1}{3} \times 4 \times 3 \times 5$ $\qquad\qquad\qquad$ $V = \dfrac{1}{3} \times 10 \times 3$

$V = 20\,\text{m}^3$ $\qquad\qquad\qquad\qquad\qquad$ $V = 10\,\text{ft}^3$

Find the volume of the following solids. Use $\pi \approx 3.14$. (DOK 2)

1.

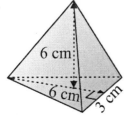

6 cm

6 cm   3 cm

2.

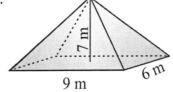

7 m

9 m   6 m

3.

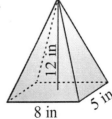

12 in

8 in   5 in

13.7   Surface Area of Cubes and Rectangular Prisms (DOK 2)

The **surface area of a solid** is the total area of all the sides of a solid.

Cube:

There are six sides on a cube. To find the surface area of a cube, find the area of one side and multiply by 6.

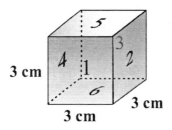

Area of each side of the cube: $3 \times 3 = 9$ cm^2

Total surface area: $9 \times 6 = 54$ cm^2

Rectangular Prism:

There are 6 sides on a rectangular prism. To find the surface area, add the areas of the six rectangular sides.

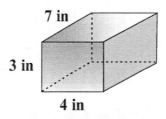

Top and Bottom

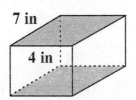

Area of top side:
7 in $\times$ 4 in $= 28$ in^2
Area of top and bottom:
28 in^2 $\times$ 2 $= 56$ in^2

Front and Back

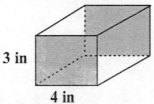

Area of front:
3 in $\times$ 4 in $= 12$ in^2
Area of front and back:
12 in^2 $\times$ 2 $= 24$ in^2

Left and Right

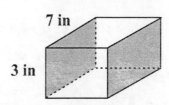

Area of left side:
3 in $\times$ 7 in $= 21$ in^2
Area of left and right:
21 in^2 $\times$ 2 $= 42$ in^2

Total surface area: 56 in^2 $+ 24$ in^2 $+ 42$ in^2 $= 122$ in^2

Find the surface area of the following cubes and prisms. (DOK 2)

1.
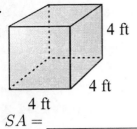
4 ft

4 ft

4 ft

SA = _____

2.
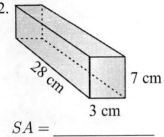
28 cm

7 cm

3 cm

SA = _____

3.
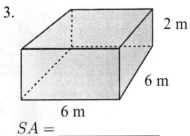
2 m

6 m

6 m

SA = _____

4.
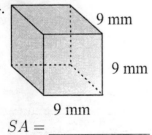
9 mm

9 mm

9 mm

SA = _____

5.
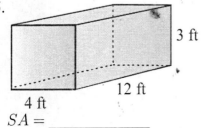
3 ft

12 ft

4 ft

SA = _____

6.
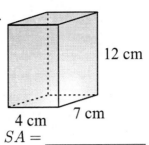
12 cm

4 cm   7 cm

SA = _____

7.
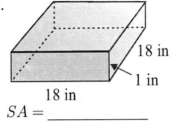
18 in

1 in

18 in

SA = _____

8.
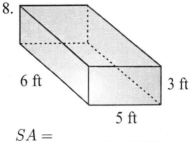
6 ft   3 ft

5 ft

SA = _____

9.
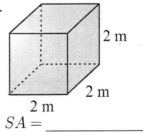
2 m

2 m

2 m

SA = _____

10.
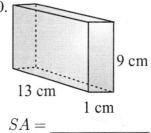
9 cm

13 cm

1 cm

SA = _____

13.8   Surface Area of Other Prisms (DOK 2)

The surface area of a prism with any polygon as its face is computed by adding the areas of the two faces and the areas of the rectangles that form the middle. For any polygon face, the number of sides will tell you how many rectangles make up the middle. To find the area of one of those rectangles, multiply the side length by the length of the prism.

Example 5: A regular hexagonal prism has length 8 in. Each of the hexagon face's sides is 4 in long, and the area of the hexagon is 41.5 in^2. What is the surface area of the prism?

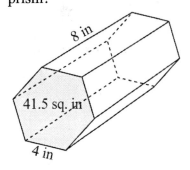

Step 1: The surface area is found by adding the areas of the two faces and the six rectangles that make up the middle of the prism.
Area of the two faces $= 41.5 \times 2 = 83$
Find the area of one middle rectangle and multiply it by the number of rectangles. We can do this because all of the rectangles are the same size.
$4 \times 8 = 32$; $32 \times 6 = 192$

Step 3: Add the areas of the faces and the middle rectangles. $192 + 83 = 275$

Answer: The surface area is 275 in^2.

Solve the following surface area problems. (DOK 2)

1. A pentagonal prism has a length of 7 cm. Its pentagon face has 5 equal side lengths of 5 cm, and its area is 43 cm^2. What is the surface area of the pentagonal prism?

2. A right triangular prism has a triangular face with sides 5, 12, and 13 in. The length of the prism is 3.5 in. What is the surface area of this triangular prism?

3. A decagonal prism has a length of 6 cm. Its decagon face has 10 equal sides of length 3 cm and an area of 69.25 cm^2. What is the surface area of the decagonal prism?

4. A hexagonal prism has a hexagon face with sides of 4.5 in and an area of 52.6 in^2. The length of the prism is 5 in long. What is the surface area of the hexagonal prism?

5. A right triangular prism has a length of 11 units. Its right triangle face has legs of 3 and 4 and a hypotenuse of 5 units. What is the surface area of the right triangular prism?

6. A heptagonal prism has a length of 12 cm. Its heptagon face has side lengths of 2, 3, 4, 3, 2, 3, and 4 cm. The area of the heptagon-shaped face is 37 cm^2. What is the surface area of the heptagonal prism?

13.9   Surface Area of Pyramids (DOK 2)

The pyramid below is made of a square base with 4 triangles on the sides.

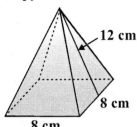

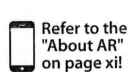

Refer to the "About AR" on page xi!

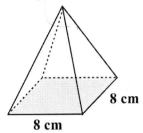

Area of square base:

$A = l \times w$

$A = 8 \times 8 = 64$ cm^2

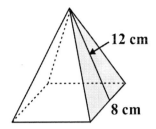

Area of sides:

Area of 1 side $= \frac{1}{2}bh$

$A = \frac{1}{2} \times 8 \times 12 = 48$ cm^2

Area of 4 sides $= 48 \times 4 = 192$ cm^2

Total surface area: $64 + 192 = 256$ cm^2

Find the total surface of the following pyramids. (DOK 2)

1.

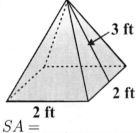

$SA =$ _____

4.

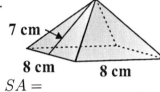

$SA =$ _____

7.

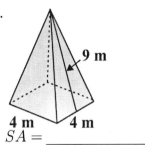

$SA =$ _____

2.

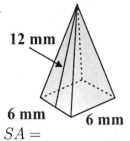

$SA =$ _____

5.

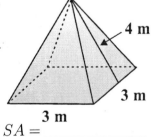

$SA =$ _____

8.

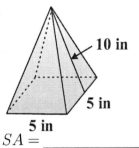

$SA =$ _____

3.

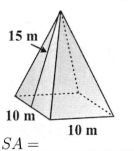

$SA =$ _____

6.

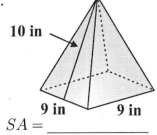

$SA =$ _____

9.

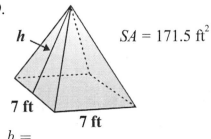

$SA = 171.5$ ft^2

$h =$ _____

13.10   Surface Area of Composite Figures (DOK 3)

Example 6:   Find the surface area of the following figure.

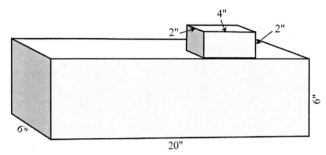

Step 1:   Find the total surface area of the top and bottom prisms.

Top: $8 \times 2 + 4 \times 2 + 8 \times 2$ Bottom: $120 \times 2 + 36 \times 2 + 120 \times 2$
       $= 40$ in^2 $= 552$ in^2

Total $= 552 + 40 = 592$ in^2

Step 2:   Subtract out the surface area that is not exposed. This is the bottom of the top figure. Remember this area is covered on both figures, so it must be subtracted out twice.
$592 - 2 \times 4 - 2 \times 4 = 592 - 8 - 8 = 576$ in^2

Find the surface areas of the figures below. (DOK 3)

1.

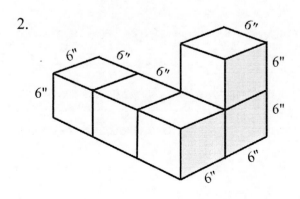

3.

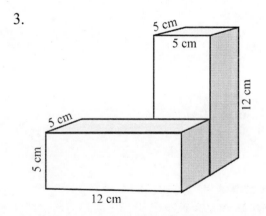

2.

4.

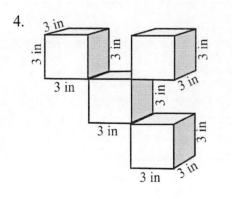

13.11   Volume and Surface Area Word Problems (DOK 3)

Example 7:   George is building a box out of cardboard. Knowing that a cube-shaped box will allow him to get the largest volume with the least amount of material, George plans to make a cube-shaped box. George has 96 sq. ft of cardboard. What will the dimensions of the box be? What volume will the box hold? Assume that George is building a closed box.

Step 1:   If George is building a cube, each of the six faces will have the same area. Find the area of one of the faces: $96 \div 6 = 16$ sq. ft.

In a cube, each face is a square, which means that both the length and width of the box are the same. What multiplied by itself is 16? $4 \times 4 = 16$. The length and width are each 4 ft.

Step 3:   Since we are working with a cube, we know the length, width, and depth are all the same measure. So the dimensions of the box are 4 ft $\times$ 4 ft $\times$ 4 ft.

Step 4:   To find the volume, multiply the length, width, and depth. 4 ft $\times$ 4 ft $\times$ 4 ft = 64 ft^3

Solve the following volume and surface area word problems. (DOK 3)

1. A swimming pool is 10 m long, 8 m wide, and 1.5 m deep. The interior of the pool will be painted with water resistant paint. How much <u>area</u> will be covered with the paint?

2. Robert wants to wrap a birthday gift for his father. The gift is in the shape of a rectangular prism that is 50 cm long, 45 cm wide, and 8 cm high. Robert has a sheet of wrapping paper that is 75 cm by 90 cm. Is the paper big enough to wrap the gift?

3. Clive cuts a 1-inch square out of each corner of an 8.5-by-11-inch piece of paper. He folds the paper into a box without a lid. What is the volume and outside surface area of the box?

4. A closet has dimensions of 12 feet by 5 feet by 6 feet. Jamie's mom uses the closet to store boxes of old toys. One box has the dimensions of 1 foot by 1 foot by 2 feet. What is the maximum number of boxes of that size that could be put in the closet?

5. A square pyramid has a base with an area of 50 cm^2 and a volume of 100 cm^3. What is the height of the pyramid?

6. An accessibility ramp shaped like a right triangular prism reaches a ledge that is 5 feet high. The ramp starts 12 feet away from the ledge, making the distance up the ramp 13 feet. The ramp will be 4.5 ft wide. All sides of the ramp will be covered with a thin layer of cement. What total area will be covered by cement?

13.12   Going Deeper into Solid Geometry (DOK 3)

Solve the surface area and volume problems below. Show your work. (DOK 3)

1. A painting crew is estimating the amount of
    paint needed to cover the bricks and the trim
    on the bottom of the office building shown
    to the right. First they must find the surface
    area of the portion of the building to be painted
    The doors and windows will not be painted.
    The larger windows measure 8 feet by 8 feet -
    there are 6 of the larger windows on each
    of three sides of the building. On the front
    of the building there are 4 larger windows and
    2 windows that measure 3 feet by 3 feet. The
    front doors will not be painted and they measure
    7 feet by 5 feet.

    The building measures 26 feet wide by 40 feet
    tall. The roof will not be painted.

    Find the area of the surface that will be painted.

2. Find the difference in volume between the box of Rice Crunchies and the box of Corn Flakies
   shown below. Then find the difference in surface area of the two boxes.

Chapter 13 Review

Find the volume and the surface area of the following solids. (DOK 2, 3)

1.

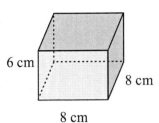

6 cm
8 cm
8 cm

$V = $ _____

$SA = $ _____

2.

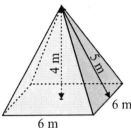

4 m
5 m
6 m
6 m

$V = $ _____

$SA = $ _____

3.

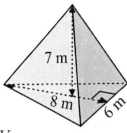

7 m
8 m
6 m

$V = $ _____

$SA = $ _____

4. If you have cubes that are two inches on each edge, how many would fit in a cube that was 18 inches on each edge?

5. It takes 27 cubic inches of water to fill the cube below. If each side of the cube is doubled, how much water is needed to fill the new cube?

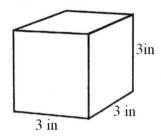

3 in
3 in
3 in

6. If a cube is 7 inches on each side, what is its volume?

7. A closed cardboard box is 20 centimeters long, 9 centimeters wide, and 12 centimeters high. What is the total surface area of the box?

8. Siena wants to build a wooden toy box with a lid. The dimensions of the toy box are 2 feet long, 5 feet wide, and 1 foot tall. How many square feet of wood will she need to construct the box?

9. How many 3-inch cubes will fit inside a larger 1 foot cube? (Figures are not drawn to scale.)

3 in
3 in
3 in

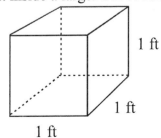

1 ft
1 ft
1 ft

Answer the following questions about cross sections and formations of solid figures. (DOK 1)

10. What three solid figures have a circle as a horizontal cross section?

11. What two solid figures can be formed by moving a square in a straight line through space?

12. What two solid figures have a triangle as a cross section?

13. A sphere is formed when what plane figure moves through space?

14. If you rotate a triangle 360°, what solid figure is formed?

15. What three solids can have a rectangle as a cross section?

Solve the following volume and surface area problems. (DOK 3)

16. An octagonal prism has length 12.75 in. The face of the prism is an octagon with area 14 in². What is the volume of the octagonal prism?

17. A hexagonal prism has length $4\frac{1}{3}$ cm. The face of the prism is a hexagon with area 18 cm².

18. A regular octagonal prism has a length of 2 units. Its octagon face has eight equal sides of length 3 units. The area of the octagon face is 43.5 units². What is the volume of the octagonal prism?

19. A regular hexagonal prism has a length of 1.5 units. Its hexagon face has six equal sides of length 2.25 units. The area of the hexagon face is 13 units². What is the volume of the hexagonal prism?

20. A triangular prism has a volume of 180 cm³ and a height of 12 cm. Find the length of a side of the triangular base if the triangle's height from that side is 6 cm.

Follow the directions for each question below. Show your work. (DOK 3)

21. Each of the colored cubes pictured to the right, are 4 inches on each side. Find the surface area of the stacks of cubes the way it is pictured.

22. Find the volume of the same stack of colored cubes.

Red	Blue	Green	Brown
Green	Red	Blue	Green
Brown	Red	Green	Blue
Blue	Brown	Red	Brown

Chapter 13 Test

1 A cube measures 3 inches on each edge. If the measure of each edge were doubled, what is the volume of the new size of the cube?

A 36 in³

B 216 in³

C 72 in³

D 1,296 in³

(DOK 2)

2 Find the volume of a rectangular prism measuring 7 cm by 9 cm by 10 cm.

A 26 cm³

B 63 cm³

C 630 cm³

D 97 cm³

(DOK 2)

3 What is the surface area of a 7 cm cube?

A 343 cm²

B 196 cm²

C 49 cm²

D 294 cm²

(DOK 2)

4 Find the surface area of the composite figure below.

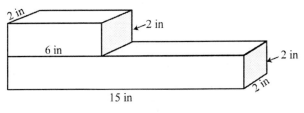

A 56 in²

B 160 in²

C 184 in²

D 368 in²

(DOK 2)

5 What shape is the vertical cross section of a cone?

A triangle

B circle

C oval

D square

(DOK 1)

6 What shape does the horizontal cross section of the figure below create?

A circle

B cylinder

C rectangle

D oval

(DOK 1)

7 A right triangular prism has a length of 9 in, and its face is a right triangle with legs of 4 in and 5 in. What is the volume of the right triangular prism?

A 90 in³

B 92 in³

C 96 in³

D 100 in³

(DOK 3)

8 A heptagonal prism has a length of 5 m. Its face is a heptagon with an area of $7\frac{2}{5}$ m². What is the volume of the heptagonal prism?

A 35 in³

B $35\frac{2}{5}$ in³

C 37 in³

D $37\frac{2}{5}$ in³

(DOK 3)

9 What is the surface area of the triangular prism shown in the figure below?

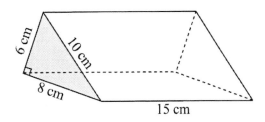

A 400 cm²
B 408 cm²
C 418 cm²
D 430 cm²

(DOK 2)

10 What is the surface area of the pentagonal prism shown in the figure below? The area of a regular pentagon with side length 6 cm is 62 cm².

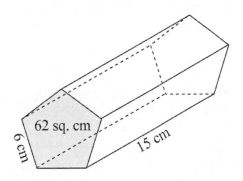

A 574 cm²
B 582 cm²
C 596 cm²
D 600 cm²

(DOK 2)

Use the figure and the information given below to answer questions 11–13.

Leah's family is going camping. Her parents bring a tent, which is in the shape of a triangular prism. The tent is 10 ft long. The triangular face is an equilateral triangle with sides of 7 ft and a height of 6 ft.

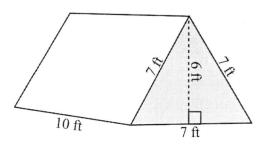

11 What is the volume of the tent?

A 105 ft³
B 155 ft³
C 210 ft³
D 230 ft³

(DOK 3)

12 What is the surface area of the tent?

A 231 ft²
B 252 ft²
C 272 ft²
D 294 ft²

(DOK 3)

13 On the next camping trip, Leah's family brings a tent that has a volume of 480 ft³. The length of the tent is 12 feet. The width is 10 feet. How tall is the tent? (The tent is shaped like a triangular prism.)

A 6 ft
B 4 ft
C 16 ft
D 8 ft

(DOK 3)

Chapter 14
Statistics

This chapter covers the following CC Grade 7 standards:

	Content Standards
Statistics and Probability	7.SP.1, 7.SP.2, 7.SP.3, 7.SP.4

14.1　Range (DOK 1)

In a given set of numbers (or data) the range is the difference between the largest and smallest values in the list. The **range** can be found by simply subtracting the smallest value from the largest value.

Example 1:　　Beaux's science test grades are 79, 83, 90, 98, and 66.
　　　　　　　　What is the range of his test grades?

　Step 1:　　Find the largest and smallest number.
　　　　　　　　The largest number is 98, and the smallest number is 66.

　Step 2:　　Subtract the smallest number from the largest number.
　　　　　　　　$98 - 66 = 32$　　　32 is the range of Beaux's test grades.

Find the range of each set of numbers. (DOK 1)

1. 51, 65, 84, 84, 21, 45, 62

2. 59, 45, 26, 41, 8, 30, 57

3. 71, 99, 88, 33, 32, 40, 77

4. 33, 100, 45, 66, 71, 88, 90

5. 45, 60, 56, 80, 90, 93, 43

6. 91, 38, 45, 33, 11, 45, 54

7. Rachel recorded the following times of the runners on Saturday: 53 minutes, 54 minutes, 47 minutes, 51 minutes, and 49 minutes. What was the range of the times Rachel collected?

8. Trece recorded the following temperatures during his chemistry experiment: 38, 43, 46, 57, 68, 79, and 29. What was the range of temperatures of Trece recorded?

9. Anna Kate is shopping for a particular mp3 player. She found the item online at 4 stores for the following prices: $45.72, $54.99, $44.88, and $51.87. What was the range of prices she found for the mp3 player she wants?

10. The 12 girls in Troop 421 sold the following numbers of boxes of cookies: 22, 40, 35, 50, 20, 18, 65, 14, 78, 190, 10, and 10. What was the range of number of boxes of cookies sold by the girls in Troop 421?

14.2   Mean (DOK 1)

The **mean** of a set of numbers is also known as the average of the set. Average is more often used to describe real life situations. To find the mean or average of a set of data, add the numbers on the list then divide by the total number of items on the list.

** Don't forget ZERO! Zero, 0, may be a number in the list. Although zero has "no" value, it must be included in the addition step and counted as a number on the list.

Example 2:   Clay had the following work hours for the week: Monday 5 hours, Tuesday 6 hours, Wednesday 0 hours, Thursday 4 hours, Friday 0 hours, Saturday 6 hours, and Sunday 0 hours. What was the mean (or average) number of hours?

   Step 1:   Find the total number of hours he worked by adding all of the hours together.
   $5 + 6 + 0 + 4 + 0 + 6 + 0 = 21$

   Step 2:   Divide the total number by the number of items in the set. The number of items is 7 (the number of days). 21 divided by 7 is 3.
   Clay's mean number of hours was 3.

Example 3:   Mr. Jones class scored the following grades on their science fair projects: 98, 96, 96, 94, 92, 88, 82, 80, 76, 74, 74, 72, 68, 68, and 60. What was the mean test score?

   Step 1:   Add up all of the grades.
   $98 + 96 + 96 + 94 + 92 + 88 + 82 + 80 + 76 + 74 + 74 + 72 + 68 + 68 + 60 = 1218$

   Step 2:   Divide the total of the grades by the number of grades in the list.
   There are 15 grades, so divide 1218 by 15.
   The mean test score is 81.2.

Find the mean for each set. Round all answers to the nearest tenth. (DOK 1)

1. 65, 70, 80, 90, 95

2. 16, 12, 8, 4, 0, 0, 10, 28, 14

3. 5, 11, 7, 9, 3

4. 4, 5, 2, 4, 5

5. 3, 6, 9, 2, 10

6. 75, 65, 85, 70, 80

7. 100, 90, 60, 0, 80, 80, 75, 85, 90, 90

8. 4.5, 6.2, 4.2, 3.2, 3.8, 4.2, 4.6, 5.4, 6.8, 2.2

9. 489, 560, 423, 550

10. $45, $50, $60, $75, $45

11. Dolly's Deli keeps track of all their sales. Here are the sales for chicken salad sandwiches for one week: 65, 76, 80, 50, 30, 88, and 50. What was the average or mean number of chicken salad sandwiches sold?

12. Deven's office tracks his incoming calls. Here is the record for his calls for this week: 16, 14, 12, 10, and 8. What was the average number of calls?

13. Tina earned $60 baby-sitting over the weekend for a total of 5 hours. What was her average rate per hour?

14. Maddie scored 18 RBIs in 10 games. What was her average RBI count per game?

15. Aaliyah drove 488 miles in 8 hours. What was her average speed?

14.3   Finding Data Missing From the Mean (DOK 2)

Example 4:   Mara knew she had an 88% average in her biology class, but she lost one of her papers. The three papers she could find had scores of 98%, 84%, and 90%. What was the score on her fourth paper?

Step 1:   Figure the total score on four papers with an 88% average. $0.88 \times 4 = 3.52$

Step 2:   Add together the scores from the three papers. $0.98 + 0.84 + 0.9 = 2.72$

Step 3:   Subtract the scores you know from the total score. $3.52 - 2.72 = 0.80$. She had 80% on her fourth paper.

Find the data missing from the following problems. (DOK 2)

1. Gabriel earns 87% on his first geography test. He wants to keep a 92% average. What does he need to get on his next test to bring his average up?

2. Rian earned $68.00 on Monday. How much money must she earn on Tuesday to have an average of $80 earned for the two days?

3. Haley, Taye, Dana, and Chris enter a contest to see who could bake the most chocolate chip cookies in an hour. They bake an average of 75 cookies. Haley bakes 55, Taye bakes 70, and Dana bakes 90. How many does Chris bake?

4. Four wrestlers make a pact to lose some weight before the competition. They lose an average of 7 pounds each over the course of 3 weeks. Carlos loses 6 pounds, Steve loses 5 pounds, and Greg loses 9 pounds. How many pounds does Wes lose?

5. Three boxes are ready for shipment. The boxes average 26 pounds each. The first box weighs 30 pounds; the second box weighs 25 pounds. How much does the third box weigh?

6. The five jockeys running in the next race average 92 pounds each. Nicole weighs 89 pounds. Jon weighs 95 pounds. Jada and Kasey weigh 90 pounds each. How much does Jordan weigh?

7. Shaneta makes three loaves of bread that weigh a total of 45 ounces. What is the average weight of each loaf?

8. Cleona makes scented candles to give away to friends. She has 2 pounds of candle wax which she melted, scented, and poured into 8 molds. What is the average weight of each candle?

9. Each basketball player has to average a minimum of 5 points a game for the next three games to stay on the team. Ben is feeling the pressure. He scored 3 points the first game and 2 points the second game. How many points does he need to score in the third game to stay on the team?

14.4   Median (DOK 1)

In a list of numbers ordered from lowest to highest, the **median** is the middle number. To find the **median**, first arrange the numbers in numerical order. If there is an odd number of items in the list, the **median** is the middle number. If there is an even number of items in the list, the **median** is the average of the two middle numbers.

Example 5: Find the median of 42, 35, 45, 37, and 41.

> Refer to the
> "About AR"
> on page xi!

Step 1: Arrange the numbers in numerical order: 35   37   41   42   45

Step 2: Find the middle number. The median is 41.

Example 6: Find the median of 14, 53, 42, 6, 14, and 46.

Step 1: Arrange the numbers in numerical order: 6   14   14 42   46   53.

Step 2: Find the average of the two middle numbers. $(14 + 42) \div 2 = 28$. The median is 28.

Circle the median in each list of numbers. (DOK 1)

1. 35, 55, 40, 30, and 45
2. 7, 2, 3, 6, 5, 1, and 8
3. 65, 42, 60, 46, and 90

4. 15, 16, 19, 25, 20
5. 75, 98, 87, 65, 82, 88, 100
6. 33, 42, 50, 22, and 19

7. 401, 758, and 254
8. 41, 23, 14, 21, and 19
9. 5, 8, 10, 13, 1, and 8

10.	11.	12.	13.	14.	15.	16.
19	9	45	52	20	8	15
14	3	32	54	21	17	40
12	10	66	19	25	13	42
15	17	55	63	18	14	32
18	6	61	20	16	22	28

Find the median in each list of numbers. (DOK 1)

17. 10, 8, 21, 14, 9, and 12
18. 43, 36, 20, and 40

19. 5, 24, 9, 18, 12, and 3
20. 48, 13, 54, 82, 90, and 7

21. 23, 21, 36, and 27
22. 9, 4, 3, 1, 6, 2, 10, and 12

23.	24.	25.	26.	27.	28.
2	11	13	75	48	22
10	22	15	62	45	19
6	25	9	60	52	15
18	28	35	52	30	43
20	10	29	80	35	34
23	23	33	50	58	28

14.5   Mode (DOK 1)

In a list of data, the **mode** is the number that appears most often. To find the mode, count the number of times that each entry appears on the list. The one that appears most is the mode. If no number appears more than any other number or if all numbers appear the same number of times, there is no mode.

Example 7:   Exam grades for a math class were as follows:
70 88 92 85 99 85 70 85 99 100 88 70 99 88 88 99 88 92 85 88

Step 1:   Count the number of times each number occurs in the list.

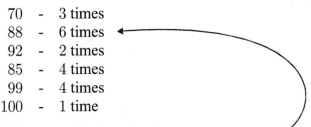

70 - 3 times
88 - 6 times
92 - 2 times
85 - 4 times
99 - 4 times
100 - 1 time

Step 2:   Find the number that occurs most often.
The mode is 88 because it is listed 6 times. No other number is listed as often.

Find the mode in each of the following lists of numbers. (DOK 1)

1.	2.	3.	4.	5.	6.	7.
88	54	21	56	64	5	12
15	42	16	67	22	4	41
88	44	15	67	22	9	45
17	56	78	19	15	8	32
18	44	21	56	14	4	16
88	44	16	67	14	7	12
17	56	21	20	22	4	12

8. 48, 32, 56, 32, 56, 48, 56

9. 12, 16, 54, 78, 16, 25, 20

10. 5, 4, 8, 3, 4, 2, 7, 8, 4, 2

11. 11, 9, 7, 11, 7, 5, 7, 7, 5

12. 84, 22, 79, 22, 87, 22, 22

13. 95, 87, 65, 94, 78, 95

14. 8, 2, 5, 4, 7, 2, 3, 6, 1

15. 89, 7, 11, 89, 17, 56

16. 15, 48, 52, 41, 8, 48

17. 22, 45, 48, 12, 22, 41, 22

18. 62, 44, 78, 62, 54, 44, 62

19. 54, 22, 54, 78, 22, 78, 22

20. 14, 17, 33, 21, 33, 17, 33

21. 65, 51, 8, 21, 8, 65, 70, 8

22. 17, 24, 13, 8, 11, 8, 15, 9

23. 51, 45, 84, 51, 65, 74, 51

24. 8, 74, 65, 15, 9, 10, 74

25. 62, 54, 2, 7, 89, 2, 7, 54, 2

14.6   Applying Measures of Central Tendency (DOK 3)

You may be asked to solve real-world problems involving measures of central tendency.

Example 8:   Aida is shopping around for the best price on a 17" computer monitor. She travels to seven stores and finds the following prices: $199, $159, $249, $329, $199, $209, and $189. When Aida goes to the eighth and final store, she finds the price for the 17" monitor is $549. Which of the measures of central tendency, mean, median, or mode, changes the most as a result of the last price Aida finds?

Step 1:   **Solve for all three measures of the seven values.**

Mean: $\dfrac{\$199 + \$159 + \$249 + \$329 + \$199 + \$209 + \$189}{7} = \$219$

Median: From least to greatest: $199, $189, $199, $199, $209, $249, $329. The 4th value = $199.
Mode: The number repeated the most is $199.

Step 2:   **Find the mean, median, and mode with the eighth value included.**

Mean: $\dfrac{\$199 + \$159 + \$249 + \$329 + \$199 + \$209 + \$189 + \$549}{8} = \$260.25$

Median: $159, $189, $199, $199, $209, $249, $329, $549. The avg. of 4th and 5th number = $204.
Mode: The number still repeated most is $199.

Answer: The measure which changed the most by adding the 8th value is the **mean**.

Follow the directions in each problem below. (DOK 3)

1. The Realty Company has the selling prices for 10 houses sold during the month of July. The following prices are given in thousands of dollars:

    176   89   525   125   107   100   525   61   75   114

    Find the mean, median, and mode of the selling prices. Which measure is most representative of the selling price of these homes? Explain.

2. A soap manufacturing company wants to know if the weight of its product is on target, meaning 4.75 oz. With that purpose in mind, a quality control technician selects 15 bars of soap from production, 5 from each shift, and finds the following weights in oz.

    1st shift: 4.76, 4.75, 4.77, 4.77, 4.74
    2nd shift: 4.72, 4.72, 4.75, 4.76, 4.73
    3rd shift: 4.76, 4.76, 4.77, 4.76, 4.76

    (A) What are the values for the measures of central tendency for the sample from each shift?
    (B) Find the mean, median, and mode for the 24 hour production sample.
    (C) Which measure is the most accurate measure of central tendency for the 24 hour production?
    (D) Find the range of values for each shift. Is the range an effective tool for drawing a conclusion in this case? Why or why not?

14.7   Comparing Two Sets of Data (DOK 2, 3)

Carefully read each situation in the problems below and answer the questions. (DOK 2, 3)

1. Tim and his stepfather went bowling. Tim's scores were 175, 142, and 181, with an average of 166. Tim's stepfather scored 134 and 187 on the first two games. What score will Tim's stepfather need in order to have the same average as Tim? Which person had greater variability in his scores?

2. Barbara and Mindy were selling bags of popcorn on two sides of the same neighborhood to raise money for the local homeless shelter. Below is a chart of the dollar amount earned by the two girls over one week.

	Mon	Tues	Wed	Thurs	Fri	Sat
Barbara	$13	$9	$12	$8	$15	$27
Mindy	$6	$14	$15	$10	$18	$33

Which girl averaged more sales per day? Who had greater variability in dollar amount earned each day?

3. Mr. Kendall bought a hobby farm after retiring and kept laying hens to start a small egg business. Below is a chart of his best egg layers and their production over a six week period.

	Week 1	Week 2	Week 3	Week 4	Week 5	Week 6
Gloria	7	6	8	7	5	6
Abigail	7	7	7	6	8	7

Using the data in the chart above,
a) What is the average weekly egg production of Gloria?
b) What is the average weekly egg production of Abigail?
c) Which chicken will give Mr. Kendall more eggs per week on average?

4. Mario and Geraldo are twin brothers who competed in a Math Throw Down, a math contest for the whole junior high school the two boys attended. Below is a chart of the results of Mario's and Geraldo's answers.

	Q 1	Q 2	Q 3	Q 4	Q 5	Q 6	Q 7	Q 8	Q9
Mario	y	n	y	y	y	n	n	y	y
Geraldo	n	y	y	y	y	y	y	n	y

For each answer they got correct, the teacher gave them a "y", meaning yes, with a value of 5 points. Each "n", meaning no, they did not get the answer correct, and received no points. Which boy scored higher and will go on to the next round?

5. Two softball teams, have each played four games so far this season. The Evergreen team scored 3, 5, 8, and 4. The Blue Skies team scored 6, 8, 2, and 12. Which team averaged a greater score and what is that average score? Which team had greater variability in his scores?

6. The heights of the girls on the Lone Pine volleyball team and the Crimson volleyball team are shown on the dot plots below.

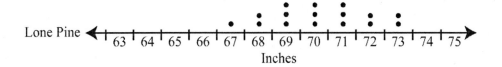

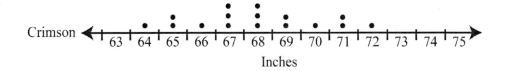

On which team are the girls generally taller? Which team has more variability?

7. Temperature highs during the month of December are shown for Georgia and New Hampshire in the dot plots below.

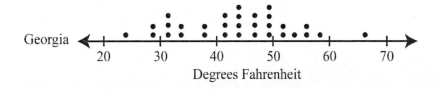

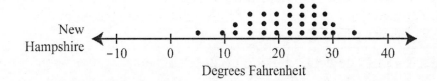

Which state is generally warmer in December? Which state has a greater variability in temperature in December?

14.8   Simple Random Sampling (DOK 2)

An important sampling technique is called **simple random sampling**. Simple random sampling is often used because it provides an excellent **representative sample** of the overall population. A representative sample is a sample that reflects (or represents) the characteristics of the whole population. There are two elements that make a sample a "simple random sample."

The first is that the sample is **unbiased**. In an unbiased sample, each member of the population is equally likely to be chosen, so the sample is representative of the population.

The second element of simple random sampling is that the observations are **independent.** In mathematics, the outcome of an event may or may not influence the outcome of a second event. If the outcome of one event does not influence the outcome of the second event these events are independent. However, if one event has an influence on the second event, the events are **dependent.**

Example 9:	Jeremy wanted to take a simple random sample of the 9th grade's movie preferences. To do so, he gave everyone in the 9th grade a ticket with a number on it, and randomly chose 15 numbers out of a hat. Then Jeremy had each of them fill out a slip of paper with their favorite movie, and he had each of them get one friend to fill out a slip with their favorite movie. Was this a simple random sample?
Solution:	No, it is not a simple random sample, because the observations are not independent. The choice of the 15 friends is dependent on the 15 students who were drawn out of the hat. Jeremy should simply draw all 30 names out of the hat.

Determine whether the following samples are simple random samples, and if not, whether they are biased or dependent. (DOK 2)

1. Henry Higgins K–8 School conducts a survey to determine children's attitudes about their parents. They get out a list of students in alphabetical order and choose every 7th name on the list to survey.

2. Million Milk tests how long it takes their milk to spoil. They use a random number generator to select 100 one-gallon bottles from among all of their products to test.

3. Reverend Nance conducts a survey on American attitudes about religion. He assigns everyone who comes to his church every Sunday a number. The reverend uses a random number generator to select 150 of them to take the survey.

4. A social scientist conducts a study on the length of time that Baltimore County newlyweds of last year dated before they married. She has the phone numbers of everyone who obtained a marriage license from Baltimore County courthouse last year, and she programs her computer to select at random 100 couples to call and survey.

Answer the following question in your own words. (DOK 2)

5. The Cuckoo Cola Company wants to gain a sense of their workers' satisfaction, so they decide to survey some of their 17,000 U.S. employees. Based on simple random sample principles, explain at least two of the key elements that should be part of the administration's design.

14.9   Representative Sampling (DOK 3)

Another important sampling technique is called **representative sampling**. Often, taking a simple random sample is not possible, either because there is no way to obtain a totally unbiased sample or a totally independent sample. The goal is still to obtain a sample as representative of the population as possible. A key element of representative sampling is **variety.**

Example 10:   Emma works for a company that produces stuffed animals at 5 factories across the Midwest. She wants to test the quality of the stuffed animals just as they come off the factory line (to eliminate the possibility of damage during shipping). Which of the following experiment designs would give the most representative sample?

(A) Go to the nearest factory and choose 100 stuffed animals at random to test.

(B) Go to each of the 5 factories and take the first 20 stuffed animals that come off the factory line.

(C) Go to each of the 5 factories and choose 20 stuffed animals at random to test.

Solution: Option (A) is a random sample that gives an excellent representation of the population of stuffed animals at that particular factory. However, Emma wants a sample that represents the population of stuffed animals at all 5 factories.

Option (C), on the other hand, is a sample that is drawn from all 5 factories. It has greater variety, so it will be more representative.

Option (B) also draws from all 5 factories, but by choosing consecutive stuffed animals, it is not an independent sample. For instance, the machines and workers may produce better stuffed animals early in the day or the week.

Option (C) is the most representative sample.

Answer the following question about representative sampling and explain your answer. (DOK 3)

1. ReallyGoodJobs.com conducts a survey to determine the attitudes of college seniors about entering the workforce. Which of these sampling techniques would give them the most representative samples?

(A) Send surveyors to 15 randomly selected college campuses across the country and ask every person who attends the school's women's soccer game to fill out a survey.

(B) Send surveyors to interview all 600 seniors at NW Maryland State University.

(C) Choose 15 randomly selected college campuses across the country and use a computer to randomly select 40 seniors from each school out of the school's student directory.

(D) Put the survey on their website and wait until 600 college seniors respond to it.

Chapter 14 Review

Find the range of each set of numbers. (DOK 1)

1. $23, 41, 76, 81, 14, 29$

2. $51, 111, 11, 1, 55, 15$

3. $99, 9, 19, 91, 119, 999$

Find the mean of each set of numbers. Round answers to the nearest tenth. (DOK 1)

4. $23, 32, 13, 12, 33, 22, 30$

5. $77, 17, 97, 79, 7, 19$

6. $34, 81, 18, 25, 66, 64$

Find the median of each set of numbers. (DOK 1)

7. $62, 17, 25, 84, 19, 26$

8. $55, 110, 115, 65, 28$

9. $875, 324, 114, 985, 645$

Find the mode of each set of numbers. (DOK 1)

10. $8, 7, 3, 2, 2, 4, 8, 7, 8, 9$

11. $21, 12, 11, 12, 22, 11, 12$

12. $18, 82, 28, 22, 82, 88, 82$

Carefully read and solve each problem below. (DOK 2, 3)

13. Chrissy went shoe shopping at several stores. She wanted one pair of black flats. The first three stores, the prices are $54, $26, and $31. Give the mean and median of the shoe prices of the first three stores.

14. Chrissy continued her shoe shopping at two more stores. Store number four charges $25 and store number five charges $36. Give the new mean and median that includes all five stores (this includes the three stores in problem 13).

15. Cindy and her friend Becky were practicing free throws everyday after school one week. The chart below shows how many baskets they made.

	Mon	Tues	Wed	Thurs	Fri
Cindy	16	15	23	28	32
Becky	12	18	14	22	26

Which girl averaged a higher score for the week? Give the name and her average.

16. Marlon, Marcus, Marion, and Emilio together averaged 86% on their math test. Marlon earned 92%, Marcus earned 78%, and Marion earned 83%. What percentage must Emilio have earned in order for the four boys to average 86%?

17. Captain Kendrick of the Starship Supernova wants to know how many of his crew of 700 are afraid of encountering intelligent aliens. Which of the following would give him a simple random sample?

    (A) Ask all of his senior officers and friends.
    (B) Tell the ship's computer to assign a number between $1 - 700$ to every crewperson and use the computer to generate random numbers to select 70 of the crew for the survey.
    (C) Get out the alphabetical crew listing and interview every 10th crew member.
    (D) Choose 10 crew members randomly from each of the 7 halls in which his crew lives.

Chapter 14 Test

1 What is the range of the following set of numbers?

14, 52, 17, 23, 65, 15

A 31
B 51
C 15
D 65

(DOK 1)

2 Find the median of the list 112, 17, 82, 128, 71, 76, and 122.

A 17
B 128
C 82
D 79

(DOK 1)

3 Find the mode of the list 31, 30, 32, 31, 30, 31, 32, 33, 30, 31, 31, 30, and 32.

A 31
B 33
C 32
D 30

(DOK 1)

4 Give the mean of the following set of numbers.

63, 25, 82, 17, 55

A 82
B 17
C 55
D 48.4

(DOK 1)

5 Earnestine received $25.00 baby-sitting in one week. How much does she need to earn the second week to average $30.00?

A $35.00
B $30.00
C $27.50
D $25.00

(DOK 2)

6 Scott mowed lawns for a total of 18 hours in one week during summer vacation and was paid a total of $252.00. He was paid by the yard, but he wanted to know how much he averaged per hour. What was Scott's average per hour?

A $15.00
B $15.50
C $14.00
D $14.50

(DOK 2)

7 Mr. and Mrs. Anderson are shopping for a new car. They want a four door car with a 6 cylinder engine. They found the following prices for the different cars that were suitable for their family:

$26,500 $23,400 $19,950
$47,800 $16,800 $18,400

What are the mean and median of the prices above?

A mean: $21,675, median: $25,475
B mean: $21,600, median: $25,400
C mean: $25,475, median: $21,675
D mean: $47,800, median: $16,800

(DOK 2)

8 There are 72 students in the school orchestra this school year. They have been invited to do a concert for the President of the United States in Washington, which will be televised nation wide. The catch is the concert will be on Homecoming weekend. The music director wants to survey the orchestra about whether or not to accept the invitation. How many orchestra members should he survey to get a representative sample?

A 12
B 24
C 36
D 42

(DOK 3)

Chapter 15
Probability

This chapter covers the following CC Grade 7 standards:

	Content Standards
Statistics and Probability	7.SP.5, 7.SP.6, 7.SP.7, 7.SP.8

15.1   Probability (DOK 2)

Probability is the chance something will happen. We express probability as a fraction, a decimal, a percent; it can also be written out in words

Example 1:   Billy has 3 red marbles, 5 while marbles, and 4 blue marbles on the floor. His cat comes along and bats one marble under the chair. What is the **probability** it is a red marble?

Step 1:   The number of red marbles, 3, will be on top of the fraction.

Step 2:   The total number of marbles, 12, will be on the bottom of the fraction. The answer may be expressed in lowest terms. $\frac{3}{12} = \frac{1}{4}$
Expressed as a decimal, $\frac{1}{4} = 0.25$, as a percent, $\frac{1}{4} = 25\%$, and written out in words, $\frac{1}{4}$ is one out of four.

Example 2:   Determine the probability that the pointer will stop on a shaded wedge or the number 1.

📱 Refer to the "About AR" on page xi!

Step 1:   Count the number of possible wedges that the spinner can stop on to satisfy the above problem. There are 5 wedges that satisfy it (4 shaded wedges and one number 1). The top number of the fraction is 5.

Step 2:   Count the total number of wedges, 7. The bottom number of the fraction is 7. The probability that the pointer will stop on a shaded wedge or the number 1 is $\frac{5}{7}$ or five out of seven.

Example 3:   Refer to the spinner in example 2. If the pointer stops on the number 7, what is the probability that it will **not** stop on 7 on the next spin?

Step 1:   Ignore the information that the pointer stopped on the number 7 on the previous spin. The probability of the next spin does not depend on the outcome of the previous spin. Simply find the probability that the spinner will not stop on 7. Remember, if P is the probability of an event occurring, $1 - P$ is the probability of an event not occurring. In this example, the probability of the spinner landing on 7 is $\frac{1}{7}$.

Step 2:   The probability that the spinner will not stop on 7 is $1 - \frac{1}{7}$ which equals $\frac{6}{7}$. The answer is $\frac{6}{7}$ or **six out of seven**.

Find the probability of the following problems. Express the answer as a percent. (DOK 2)

1. A computer chooses a random number between 1 and 50. What is the probability of your guessing the same number that the computer chooses in 1 try?

2. There are 24 candy-coated chocolate pieces in a bag. Eight have defects in the coating that can be seen only with close inspection. What is the probability of pulling out a defective piece without looking?

3. Seven sisters have to choose which day each will wash the dishes. They put equal-sized pieces of paper each labeled with a day of the week in a hat. What is the probability that the first sister who draws will choose a weekend day?

4. For his garden, Clay has a mixture of 12 white corn seeds, 24 yellow corn seeds, and 16 bicolor corn seeds. If he reaches for a seed without looking, what is the probability that Clay will plant a bicolor corn seed first?

5. Mom just got a new department store credit card in the mail. What is the probability that the last digit is an odd number?

6. Alex has a paper bag of cookies that includes 8 chocolate chip, 4 peanut butter, 6 butterscotch chip, and 12 ginger. Without looking, his friend John reaches in the bag for a cookie. What is the probability that the cookie is peanut butter?

7. An umpire at a little league baseball game has 14 balls in his pockets. Five of the balls are brand A, 6 are brand B, and 3 are brand C. What is the probability that the next ball he throws is a brand C ball?

8. What is the probability that the spinner arrow will land on an even number?

9. Using the spinner above, what is the probability that the spinner will not stop on a shaded wedge or on the number 2 on the third spin?

10. A company is offering 1 grand prize, 3 second place prizes, and 25 third place prizes based on a random drawing of contest entries. If you entered one of the 500 total entries, what is the probability you will win a third place prize?

11. In the contest problem above, what is the probability that you will win the grand prize or a second place prize?

12. A box of a dozen doughnuts has 3 lemon cream-filled, 5 chocolate cream-filled, and 4 vanilla cream-filled. What is the probability of picking a lemon cream-filled?

15.2   Independent and Dependent Events (DOK 2)

In mathematics, the outcome of an event may or may not influence the outcome of a second event. If the outcome of one event does not influence the outcome of the second event, these events are **independent**. However, if one event has an influence on the second event, the events are **dependent**. When someone needs to determine the probability of two events occurring, he or she will need to use an equation. These equations will change depending on whether the events are independent or dependent in relation to each other. When finding the probability of two **independent** events, multiply the probability of each favorable outcome together.

Example 4:   One bag of marbles contains 1 white, 1 yellow, 2 blue, and 3 orange marbles. A second bag of marbles contains 2 white, 3 yellow, 1 blue, and 2 orange marbles. What is the probability of drawing a blue marble from each bag?

Solution: Probability of favorable outcomes

Bag 1: $\dfrac{2}{7}$

Bag 2: $\dfrac{1}{8}$

Probability of a blue marble from each bag: $\dfrac{2}{7} \times \dfrac{1}{8} = \dfrac{2}{56} = \dfrac{1}{28}$

In order to find the probability of two **dependent** events, you will need to use a different set of rules. For the first event, you must divide the number of favorable outcomes by the number of possible outcomes. For the second event, you must subtract one from the number of favorable outcomes **only if** the favorable outcome is the **same**. However, you must subtract one from the number of total possible outcomes. Finally, you must multiply the probability for event one by the probability for event two.

Example 5:   One bag of marbles contains 3 red, 4 green, 7 black, and 2 yellow marbles. What is the probability of drawing a green marble, removing it from the bag, and then drawing another green marble?

	Favorable Outcomes	Total Possible Outcomes
Draw 1	4	16
Draw 2	3	15
Draw 1 × Draw 2	12	240

Answer: $\dfrac{12}{240}$ or $\dfrac{1}{20}$

Example 6:   Using the same bag of marbles, what is the probability of drawing a red marble and then drawing a black marble?

	Favorable Outcomes	Total Possible Outcomes
Draw 1	3	16
Draw 2	7	15
Draw 1 × Draw 2	21	240

Answer $\dfrac{21}{240}$ or $\dfrac{7}{80}$

HINT: When asked to find the probability of two events that are occurring, there are two key words to look for: "or" and "and." If a problem asks for the probability of one event **or** another event occurring, then you add the probabilities of the two events together. If a problem asks for the probability of one event **and** another event occurring, then you multiply the probabilities of the two events together. You multiply because the chance of having two different events happen is less than the chance of only one event happening, and when you multiply two fractions, you always get a smaller fraction.

Find the probability of the following problems. Express the answer as a fraction. (DOK 2)

1. Prithi has two boxes. Box 1 contains 3 red, 2 silver, 4 gold, and 2 blue combs. She also has a second box containing 1 black and 1 clear brush. What is the probability that Prithi selected a red comb from box 1 and a black brush from box 2?

2. Steve Marduke had two spinners in front of him. The first one was numbered 1–6, and the second was numbered 1–3. If Steve spins each spinner once, what is the probability that the first spinner will show an odd number and the second spinner will show a "1"?

3. Carrie McCallister flipped a coin twice and got heads both times. What is the probability that Carrie will get tails the third time she flips the coin?

4. Artie Drake turns a spinner which is evenly divided into 11 sections numbered 1–11. On the first spin, Artie's pointer lands on "8." What is the probability that the spinner lands on an even number the second time he turns the spinner?

5. Leanne Davis played a game with a street entertainer. In this game, a ball was placed under one of three coconut halves. The vendor shifted the coconut halves so quickly that Leanne could no longer tell which coconut half contained the ball. She selected one and missed. The entertainer then shifted all three around once more and asked Leanne to pick again. What is the probability that Leanne will select the coconut half containing the ball?

6. What is the probability that Jane Robelot reaches into a bag containing 1 daffodil and 2 gladiola bulbs and pulls out a daffodil bulb, and then reaches into a second bag containing 6 tulip, 3 lily, and 2 gladiola bulbs and pulls out a lily bulb?

7. Terrell cast his line into a pond containing 7 catfish, 8 bream, 3 trout, and 6 northern pike. He immediately caught a bream. What are the chances that Terrell will catch a second bream the next time he casts his line?

8. Gloria Quintero entered a contest in which the person who draws his or her initials out of a box containing all 26 letters of the alphabet wins the grand prize. Gloria reaches in, draws a "G," keeps it, then draws another letter. What is the probability that Gloria will next draw a "Q"?

9. Vince Macaluso is pulling two socks out of a washing machine in the dark. The washing machine contains three tan, one white, and two black socks. If Vince reaches in and pulls out the socks one at a time, what is the probability that he will pull out two tan socks on his first two tries?

10. John Salome has a bag containing 2 yellow plums, 2 red plums, and 3 purple plums. What is the probability that he reaches in without looking and pulls out a yellow plum and eats it, then reaches in again without looking and pulls out a red plum to eat?

15.3   More Probability (DOK 2)

Example 7:   You have a cube with one number, 1, 2, 3, 4, 5 and 6 painted on each face of the cube. What is the probability that if you throw the cube 3 times, you will get the number 2 each time?

If you roll the cube once, you have a 1 in 6 chance of getting the number 2. If you roll the cube a second time, you again have a 1 in 6 chance of getting the number 2. If you roll the cube a third time, you again have a 1 in 6 chance of getting the number 2. The probability of rolling the number 2 three times in a row is:

$$\frac{1}{6} \times \frac{1}{6} \times \frac{1}{6} = \frac{1}{216}$$

Find the probability that each of the following events will occur. (DOK 2)

There are 10 balls in a box, each with a different digit on it: 0, 1, 2, 3, 4, 5, 6, 7, 8, & 9. A ball is chosen at random and <u>then put back in the box.</u>

1. What is the probability that if you picked out a ball 3 times, you would get number 7 each time?

2. What is the probability you would pick a ball with 5, then 9, and then 3?

3. What is the probability that if you picked out a ball 4 times, you would always get an odd number?

4. A couple has 4 children ages 9, 6, 4, and 1. What is the probability that they are all girls?

There are 26 letters in the alphabet allowing a different letter to be on each of 26 cards. The cards are shuffled. After each card is chosen at random, it is put back in the stack of cards, and the cards are shuffled again.

5. What is the probability that when you pick 3 cards, you would draw first a "y," then and "e," and then an "s"?

6. What is the probability that you would draw 4 cards and get the letter "z" each time?

7. What is the probability that you draw twice and get a letter in the word "random" both times?

8. If you flipped a coin 3 times, what is the probability you would get heads every time?

9. Marie is clueless about 4 of her multiple-choice answers. The possible answers are A, B, C, D, E, or F. What is the probability that she will guess all four answers correctly?

15.4   Simulations (DOK 3)

A **simulation** is usually generated by a computer program. It automatically produces the results of an experiment. To find probabilities, the simulation generates the results from a series of trials. The probabilities that are found from simulations are experimental and are not always accurate.

Example 8:   The chart below represents a computer simulation. It shows the frequencies of the results of flipping two coins. The two coins were flipped at the same time 100 times.

Outcome	TT	TH	HT	HH
Frequency	23	35	23	19

Find the theoretical probability of flipping one tail and one head, and find the experimental probability of flipping one tail and one head based on the computer simulation, then compare the two values.

Step 1:   Find the theoretical probability. The probability of flipping a tail with coin one is $\frac{1}{2}$, and the probability of flipping a head with coin two is $\frac{1}{2}$. To find the probability of flipping a tail with coin one and flipping a head with coin two, you must multiply the two probabilities together, $\frac{1}{2} \times \frac{1}{2} = \frac{1}{4}$. The probability of flipping a head with coin one and a tail with coin two is $\frac{1}{2} \times \frac{1}{2} = \frac{1}{4}$. Since it does not matter which coin is tails and which is heads, add the two probabilities together.

$$\frac{1}{4} + \frac{1}{4} = \frac{1}{2}$$

The theoretical probability is 50%.

Step 2:   Find the experimental probability. The frequency of TH is 35, so out of 100 flips, the probability is $\frac{35}{100}$. The frequency of HT is 23, so out of 100 flips, the probability is $\frac{23}{100}$. To find the theoretical probability of flipping one head and one tail, you need to add the two probabilities together.

$$\frac{35}{100} + \frac{23}{100} = \frac{58}{100} = \frac{29}{50}$$

The experimental probability based on the simulation is 58%.

Step 3:   The difference between the theoretical probability and the experimental probability is 8%. Eight percent is not a huge difference. Since the two values are not too far apart, this means that the computer accurately simulates tossing two coins.

Use the simulations to find your answers. (DOK 3)

1. A computer program simulated tossing three coins 500 times. The results are shown below.

HHH	50		HTT	66
HTH	76		THT	57
HHT	62		TTH	69
THH	64		TTT	56

(A) Based on the computer simulation, what is the experimental probability of tossing two heads and a tail?

(B) What is the theoretical probability of tossing two heads and one tail?

(C) Based on the computer simulation, what is the experimental probability of tossing three tails?

(D) What is the theoretical probability of tossing three tails?

(E) Compare your answers from part (A) with part (B) and compare your answer from part (C) with part (D). Based on this comparison, is this chart an accurate simulation of tossing three coins?

2. Below is a computer simulation of rolling one six-sided cube 50 times.

Outcome	1	2	3	4	5	6
Frequency	8	6	13	11	5	7

(A) What is the theoretical probability of rolling a 3 or a 4?

(B) Calculate the theoretical probability of rolling a 6.

(C) Determine what the experimental probability of rolling a six based on the simulation.

(D) Compare the theoretical and experimental probabilities of rolling a six from parts (B) and (C), what are your conclusions?

15.5   Tree Diagrams (DOK 2)

Drawing a tree diagram is another method of determining the probability of events occurring.

Example 9:   If you toss two six-sided numbered cubes that have 1, 2, 3, 4, 5, or 6 on each side, what is the probability you will get two cubes that add up to 9? One way to determine the probability is to make a tree diagram.

Cube 1	Cube 2	Cube 1 plus Cube 2
1	1	2
	2	3
	3	4
	4	5
	5	6
	6	7
2	1	3
	2	4
	3	5
	4	6
	5	7
	6	8
3	1	4
	2	5
	3	6
	4	7
	5	8
	6	⑨
4	1	5
	2	6
	3	7
	4	8
	5	⑨
	6	10
5	1	6
	2	7
	3	8
	4	⑨
	5	10
	6	11
6	1	7
	2	8
	3	⑨
	4	10
	5	11
	6	12

Alternative method

Write down all of the numbers on both cubes which would add up to 9.

Cube 1	Cube 2
4	5
5	4
6	3
3	6

Numerator = 4 combinations

For denominator: Multiply the number of sides on one cube times the number of sides on the other cube.

$6 \times 6 = 36$

Numerator:
Denominator: $\dfrac{4}{36} = \dfrac{1}{9}$

There are 36 possible ways the cubes could land. Out of those 36 ways, the two cubes add up to 9 only 4 times. The probability you will get two cubes that add up to 9 is $\dfrac{4}{36}$ or $\dfrac{1}{9}$.

Read each of the problems below. Then answer the questions. (DOK 2)

1. Jake has a spinner. The spinner is divided into eight equal regions numbered 1–8. In two spins, what is the probability that the numbers added together will equal 12?

2. Charlie and Libby each spin one spinner one time. The spinner is divided into 5 equal regions numbered 1–5. What is the probability that these two spins added together would equal 7?

3. Gail spins a spinner twice. The spinner is divided into 9 equal regions numbered 1–9. In two spins, what is the probability that the difference between the two numbers will equal 4?

4. Diedra throws two 10-sided numbered polyhedrons. What is the probability that the difference between the two numbers will equal 7?

5. Cameron throws two six-sided numbered cubes. What is the probability that the difference between the two numbers will equal 3?

6. Tesla spins one spinner twice. The spinner is divided into 11 equal regions numbered 1–11. What is the probability that the two numbers added together will equal 11?

7. Samantha decides to roll two five-sided numbered prisms. What is the probability that the two numbers added together will equal 4?

8. Mary Ellen spins a spinner twice. The spinner is divided into 7 equal regions numbered 1–7. What is the probability that the product of the two numbers equals 10?

9. Conner decides to roll two six-sided numbered cubes. What is the probability that the product of the two numbers equals 4?

10. Tabitha spins one spinner twice. The spinner is divided into 9 equal regions numbered 1–9. What is the probability that the sum of the two numbers equals 10?

11. Darnell decides to roll two 15-sided numbered polyhedrons. What is the probability that the difference between the two numbers is 13?

12. Inez spins one spinner twice. The spinner is divided into 12 equal regions numbered 1–12. What is the probability that the sum of two numbers equals 10?

13. Gina spins one spinner twice. The spinner is divided into 8 equal regions numbered 1–8. What is the probability that the two numbers added together equals 9?

14. Celia rolls two six-sided numbered cubes. What is the probability that the difference between the two numbers is 2?

15. Brett spins one spinner twice. The spinner is divided into 4 equal regions numbered 1–4. What is the probability that the difference between the two numbers will be 3?

15.6   Probability Models (DOK 3)

Probability models allow us to say how likely an event is. Uniform probability models are developed from the assumption that all outcomes are equally likely. Observed probability models are developed based on observed frequencies.

Example 10:   (A) Assuming that there is an equal likelihood for any number to be rolled on a die, what is the probability that you will roll a 2?

(B) Now suppose that out of several trials, a 2 was rolled 40% of the time and each of the other numbers were rolled an equal number of times. Based on this, what is the observed probability model?

(C) Why might the two probability models be different?

Step 1:   The probability that you will roll a 2 on a die is 1 out of 6, or $\frac{1}{6} = 0.17$. This means you should roll a 2 17% of the time.

Step 2:   A 2 will be rolled 40% of the time. The other 60% is split equally among the other 5 numbers. Therefore, the observed probability model will be that a 2 is rolled 40% of the time and the other numbers are each rolled 12% of the time.

Step 3:   There are two main reasons why an observed probability model may be different from a uniform probability model. The first reason is that perhaps only a small number of trials were used to find the observed probabilities. If a die was rolled only 10 times, one number could have been rolled 4 times and another number not rolled at all. There is not enough evidence to conclude that the uniform probability model is incorrect. The second reason that the probability models may be different is that the probabilities of each outcome are not equal. Maybe the die that was used is weighted to produce a greater chance of landing on a certain number.

Example 11:   (A) Assuming that there is an equal likelihood that a girl or a boy will be selected from Miss May's class, what is the probability that a boy will be chosen.
(B) Now suppose that out of 10 trials, a boy was chosen 7 times. Using this data, what is the approximate probability that a boy will be chosen from the class?
(C) Why might the two probabilities be different?

Step 1:   If the chances for picking a boy or a girl are equal, the probability of choosing a boy is 0.5, or 50%.

Step 2:   A boy was chosen 7 out of 10 times. That means a boy was chosen 70% of the time. Therefore, the approximate probability of choosing a boy is 70%.

Step 3:   The two probabilities might be different due to a small number of trials. If more students were chosen, perhaps the approximate probability would've been closer to 50%. Another reason the probabilities might differ is the proportion of boys in Miss May's class. There may be more boys than girls in the class, which would explain why a boy was chosen more than 50% of the time.

Solving the following probability model problems. (DOK 3)

1. In a drawer, there are blue, yellow, green, orange, and purple socks.

   (A) Assuming there is an equal number of each color of sock, what is the probability that Josie will randomly pull out a yellow sock?
   (B) Suppose that out of 150 times, she pulled out a yellow sock 12 times. Knowing this, what is the approximate, or observed, probability that a yellow sock will be chosen?
   (C) Why might the two probabilities be different?

2. Barnum's animal crackers currently features 20 different animal shapes.

   (A) Assuming an equal number of each animal shape, what is the likelihood that Benny will pull out a rhinoceros from a bag of Barnum's animal crackers?
   (B) Suppose that out of 40 times, he pulled out a rhinoceros 3 times. What is the approximate, or observed, probability of picking a rhinoceros from the bag?
   (C) Why might the two probabilities be different?

3. In a game, if Sarah rolls a 5 or 6 on a die, she will receive a special advantage over her opponent.

   (A) Assuming an equal chance to roll every number on the die, what is the probability that Sarah will roll a 5 or 6?
   (B) Suppose that out of 500 times, she rolled 5 or 6 215 times. What is the approximate probability of rolling a 5 or 6?
   (C) Based on the information given in the previous part, does it appear that rolling each number is equally likely?

4. When counting the number of each color that was in a bag of skittles, Kellie found the following distribution:

Red	Orange	Yellow	Green	Purple
10	8	4	25	19

Skittles claims that there is an equal distribution of colors in each bag. Based on Kellie's findings, do the colors appear to be equally distributed?

5. After rolling a die 250 times, Casey recorded the following distribution of outcomes:

One	Two	Three	Four	Five	Six
40	42	41	42	42	43

Based on Casey's results, does it appear that each outcome is equally likely?

15.7   Predictions Using Probabilities (DOK 3)

Once you know the probability of an event, you can use that probability to make predictions about future events.

Example 12:   Slugger Jones has a 0.475 batting average this season, meaning that he gets a hit 47.5% of his official at-bats (excluding walks, hit by pitch, etc). In the double-header yesterday, he had 11 official at-bats. How many hits would you have expected him to collect?

Answer: Multiply the probability by the number of total opportunities. $0.475 * 11 = 5.225$ We would expect him to have gotten 5 hits in 11 at-bats.

Example 13:   Farmer Jacob owns so many sheep that he has 5 assistant shepherds. He wants to check on the composition of his flock, so he selects 40 sheep at random to see what they look like. Striped sheep are the most valuable. The results of his survey are below:

Type of sheep	White	Black	Striped	Speckled
Number of sheep	15	12	7	6

If one sheep is chosen at random, what is the probability that it is striped? If Farmer Jacob has 750 sheep total, how many of his sheep will probably be striped?

Step 1:   The probability of one sheep being striped is
$$\frac{\text{The number of striped sheep}}{\text{The total number of sheep}} = \frac{7}{40} = 17.5\%$$

Step 2:   Multiply the probability of one sheep being striped by the total number of sheep in the flock:
$0.175 \times 750 = 131.25$
He most likely has 131 striped sheep.

Example 14:   Mr. Gallop takes a poll of a randomly selected portion of the 10th grade class about their favorite baseball team. He makes a chart with the top 5 responses:

Top 5 Responses Among 10th Graders to the Question, "Who is Your Favorite Baseball Team?"

Favorite Team	Baltimore	Washington, DC	New York	Chicago	Boston
# of Responses	24	18	7	6	5

Mr. Manyin reads this chart and counts that there are 63 responses listed. He concludes that there is a $\frac{7}{63} = \frac{1}{9}$ probability that a randomly selected student would be a New York fan.
Therefore, he says, out of the 432 students in the 10th grade, $432 \times \frac{1}{9} = 48$ are likely to be New York fans. Why is his conclusion invalid?

Answer: The chart only gives the Top 5 responses about 10th graders to Mr. Gallop's question, not every response. Therefore, we do not know how many total students were polled, and we cannot calculate probabilities or make predictions.

Use the probabilities in the questions below to make predictions, if possible. If it is not possible to make predictions, explain why. (DOK 3)

1. Mrs. Polley's oven burns about 6% of the cookies in a batch, on average. She makes 400 cookies for her daughter Lynn's wedding. How many cookies should she expect her oven to burn?

2. Christie Jo is on the Prom Planning Panel (PPP). She does a random survey of 30 female classmates to find out their plans for prom. The results are shown below.

Prom Plans	Have a Date	Hope to Have a Date	Attending without Date	Not Attending
# of Females	5	13	3	9

What is the probability that a female student selected at random is planning to attend? If there are 340 female students eligible to go to the prom, how many should Christie Jo expect to attend (based on this survey)?

3. Prisci surveyed a random sample of her classmates about their favorite type of ice cream. The results are shown below.

Favorite Types of Ice Cream at Grosami High School

Favorite Ice Cream*	Vanilla (incl. French Vanilla)	Chocolate	Strawberry	Chocolate Chip
# of Responses	24	8	5	3

* lists only ice cream flavors that received 3 or more responses

According to the survey results, what is the probability that a randomly selected student would prefer chocolate? If there are 80 students total in Prisci's class, how many of them would she predict prefer chocolate?

4. Van uses the SureFire cell phone company. They claim to have a dropped call rate of 5% in his area. If Van plans on making 60 phone calls this month, how many dropped calls should he expect?

5. Peggy takes a lot of photographs, and she prefers the quality of old-style film cameras. She buys rolls of film that take 60 pictures each. The table below shows the number of photos from each roll that Peggy rejects from putting in a photo album.

Peggy's Rejected Photos (rolls of 60)

Film Roll	1	2	3	4	5	6	7
Rejected Photos	3	8	2	0	4	5	5

In a year, Peggy uses 40 rolls, or 2400 photos. Based on these 7 rolls, how many photos would you expect her to reject? Round to the nearest whole number.

15.8   Equally Likely Versus Equally Probable (DOK 2)

Since probability is often written as a fraction, decimal, or percent, it is always a number between 0 and 1. The likelihood of an event occurring can be described based on probability as follows:

If the probability is:	The event will:
1.0	certainly or definitely occur
0.75	most likely occur
0.5	equally likely occur or not occur
0.25	not likely to occur
0	not occur

If probability is written as a fraction, change it to a decimal to determine the likelihood of the event happening.

Use the above descriptions to answer the problems below. (DOK 2)

1. Steve is doing a science experiment on the probability of rolling a regular 6-sided die. He has figured out that he will roll a 1, 2, 3, 4, 5, or 6. Help Steve complete the chart below.

Event	Probability (fraction and decimal)	How likely is this to happen?
Rolling an even number		
Rolling a factor of 6		
Rolling a whole number		
Rolling an odd number		
Rolling a 7		
Rolling a multiple of 2		

2. Steve's science fair question is "Which is more likely to occur when rolling a die 100 times: an even number or a factor of 6?" Complete Steve's hypothesis. Based on the mathematical probability of each event, I hypothesize that a factor of 6 is _____ to be rolled than an even number.

Chapter 15 Review

Carefully read and solve each problem. (DOK 2)

1. There are 50 students in the school orchestra in the following sections:

25	string section
15	woodwind
5	percussion
5	brass

   One student will be chosen at random to present the orchestra director with an award. What is the probability the student will be from the woodwind section?

2. Fluffy's cat treat box contains 6 chicken-flavored treats, 5 beef-flavored treats, and 7 fish-flavored treats. If Fluffy's owner reaches in the box without looking and chooses one treat, what is the probability that Fluffy will get a chicken-flavored treat?

3. The spinner stops on the number 5 on the first spin. What is the probability that it will not stop on 5 on the second spin?

4. Sherri turns the spinner on the board below 3 times. What is the probability that the pointer always lands on a shaded number?

5. Three cakes are sliced into 20 pieces each. Each cake contains 1 gold ring. What is the probability that one person who eats one piece of cake from each of the 3 cakes will find 3 gold rings?

6. Brianna tosses a coin 4 times. What is the probability she gets all tails?

7. A box of a dozen doughnuts has 3 lemon cream-filled, 5 chocolate cream-filled, and 4 vanilla cream-filled. If the doughnuts look identical, what is the probability that if you pick a doughnut at random, it will be chocolate cream-filled?

8. Erica gets a new credit card in the mail. What is the probability that the last four digits are all 5's?

9. There are 26 letters in the alphabet. What is the probability that the first two letters of your new license plate will be your initials?

10. Mary has 4 green mints and 8 white mints the same size in her pocket. If she pulls out one, what is the probability it will be green?

Read the following, and answer questions 11–14. (DOK 2)

There are 9 slips of paper in a hat, each with a number from 1 to 9. The numbers correspond to a group of students who must answer a question when the number for their group is drawn. Each time a number is drawn, the number is put back in the hat.

11. What is the probability that the number 6 will be drawn twice in a row?

12. What is the probability that the first 5 numbers drawn will be odd numbers?

13. What is the probability that the second, third, and fourth numbers drawn will be even numbers?

14. What is the probability that the first five times a number is drawn it will be the number 5?

Carefully read each questions and solve. (DOK 3)

15. Healthy Foods Grocery Store makes fruit salads according to the following distribution:

Strawberry	Blueberry	Pineapple	Grape	Cantaloupe	Watermelon
12%	15%	18%	15%	20%	20%

Henry bought a fruit salad and counted the amount of each type of fruit. There were a total of 120 pieces of fruit in the salad. He came up with the following distribution:

Strawberry	Blueberry	Pineapple	Grape	Cantaloupe	Watermelon
14	18	22	19	24	23

Based on Henry's findings, does it appear that Healthy Foods' distribution is accurate?

16. Drees works in a factory that makes combs. The following chart shows the number of defective combs per batch (1 batch is 500 combs):

Batch Number	1	2	3	4
Number of Defective Combs	34	23	33	26

If the factory produces 50 batches a month, how many defective combs should Drees expect them to produce?

17. What is the likelihood of the 4th of July being on a Sunday? Use probability to explain your answer.

18. How likely is Thanksgiving Day to be on a Sunday? How do you know?

Chapter 15 Test

1 Mrs. O'Brien was looking for a way to have fun the day before Spring Break. She decided to "Spin the Day Away". Looking at the spinner below, what is the probability of the first spin landing on either art or cleaning? Express your answer in fraction form.

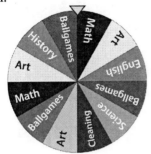

A $\frac{1}{3}$

B 33%

C $\frac{1}{4}$

D 0.25

(DOK 2)

2 A basket of kittens just woke up from their nap. Looking at the basket below, what is the probability that a kitten with only black fur will be the first one to jump out of the basket? Express your answer in decimal form, rounding to the nearest hundredth.

A 0.25
B 33%
C 25%
D 0.33

(DOK 2)

3 In a cookie jar containing 6 chocolate chip cookies, 8 wafer cookies, and 10 double chocolate cookies, what is the probability that a person who isn't looking as they pull out one cookie, will get a chocolate chip cookie? Express your answer in percent form.

A 20%

B 15%

C 25%

D 33%

(DOK 2)

4 Rita and Rose were sharing a bag of jelly beans. The bag contained 52 orange jelly beans, 63 lime ones, 72 cherry ones, 28 lemon ones, and 65 pink ones of an unknown flavor. If Rita closes her eyes and pulls out the first jelly bean, what is the probability that the jelly bean she picks is a lemon one? Express your answer in percentage form.

A 28%

B 14%

C 10%

D 65%

(DOK 2)

5 Carrie bought a large basket of 60 apples. When she got them home, she found 4 of the apples were rotten. If she goes back and buys 200 more apples, about how many rotten apples should she expect?

  A 8

  B 13

  C 20

  D 24

(DOK 2)

6 Lilly won first place in a math game in Mrs. Eberstadt's class, so she gets to pull a prize from the prize box. She doesn't know what she wants, so she is going to choose randomly. There 7 different prizes. Assuming there is an equal number of each prize in the box, what is the probability that she will pull out a gift card for a fast food restaurant?

  A 16.7%
  B 12.5%
  C 14.3%
  D 20%

(DOK 2)

7 Suppose that out of 30 times, a gift card for a fast food restaurant was randomly pulled from the prize box 7 times. What is the approximate probability that this gift card will be chosen from the box?

  A 14.3%
  B 17%
  C 20%
  D 23.3%

(DOK 3)

8 Carrie's team has won 8 of the last 9 ball games. How likely is it that they will win again based on this record?

  A very likely
  B not possible
  C unlikely
  D impossible to predict

(DOK 2)

9 On a regular 6-sided die which is more likely: rolling an odd number or an even number?

  A rolling an odd number
  B rolling an even number
  C the events are equally likely
  D neither are possible

(DOK 2)

10 A candy factory does a study to see how many candies are not usable due to an aging machine on the assembly line. The following chart shows the number of unusable candies in each 500 piece run.

Run #1	Run #2	Run #3	Run #4	Run #5
22	23	24	22	23

If the factory produces 2,000 runs of 500 pieces each day, how many unusable candies can they expect each day?

  A 228, 000
  B 45, 600
  C 54, 600
  D 11, 400

(DOK 3)

Chapter 16
How to Answer Constructed-Response Questions

16.1   Writing Short Answers

At times, you will need to do more than fill in blanks or make lists. Be sure to answer the question clearly. Look at this example below:

Example 1:   Mark gets a weekly allowance of $10.00. Mark decides to save his weekly allowance for four weeks so that he can purchase a stereo system for his room. The stereo system costs exactly $40.00 with tax included. When Mark goes to buy the stereo, he sees that the stereo is on sale. The sale includes the total price of the stereo with tax included. If Mark has exactly $10.40 left when he leaves the store, how much did Mark pay for the stereo? How much was the percentage discount on the stereo system?

Step 1:   Find out the total amount of money Mark has before he bought the stereo.

10.00×4 weeks $= \$40.00$

Step 2:   Find out how much money Mark paid for the stereo. Subtract what he has left from what he began with.

$\$40.00 - \$10.40x = \$29.60$

Mark paid $29.60 for the stereo.

Step 3:   Set up a proportional relationship of the original cost to the amount left over after the purchase.

$$\frac{\$40.00}{\$10.40} = \frac{100\%}{x}$$

Step 4:   Cross multiply to solve the proportion for x.

$\$40.00x = 1 \times \$10.40$

$\$40.00x = \$10.40$

$x = \dfrac{\$10.40}{\$40.00}$

$x = 0.26 = 26\%$

Answer: Mark paid $29.60 for the stereo, and the percentage discount on the stereo was 26%.

16.2   Writing Open-Ended Answers

Read the Question Carefully

First of all, read the question carefully. Make sure you understand what it is asking. If you need help, ask your teacher.

Write Clearly

Answer each question in a clear way. If a question asks when something happened, be sure to talk about the sequence of events. If it asks you to compare two things, be sure to talk about how they are similar.

Being clear also means that your writing must be correct. Be sure to check your spelling. Also look for any other mistakes. If you find an error, erase it completely or cross it out. Write what you really wanted to say right above it or next to it.

Use Neat Handwriting

Make sure that the people reading your answer can tell what it is. Write in a neat way that others can read. Label your answer "answer:".

Here is an example. Read the question carefully. Look at exactly what it asks you to do. Then, you will practice answering the question. You will also see how the best answer looks.

Example 2:   Once the pool was built, the city officials noticed that the builders left a big oak tree too close to the pool. It dropped leaves, twigs, and bugs into the water constantly. Keeping the pool clean was a problem. Finding out if there was enough room to cut down the tree was also a problem. City officials measured the shadow of the tree and found that its shadow was 36 feet long at the same time that an 8 foot pole standing straight up had a shadow of 3 feet. Write a proportion and then solve it to find out how tall the tree is. Show your work.

A Set up the proportion.

B Solve for the height of the tree.

Good Answer:

A $\dfrac{36 \text{ feet}}{3 \text{ feet}} = \dfrac{x}{8 \text{ feet}}$

B 96 feet

Better Answer:

A shadow length   height

$\dfrac{36 \text{ feet}}{3 \text{ feet}} = \dfrac{x}{8 \text{ feet}}$

B $36\,(8) = 3x$

$x = 96$ feet

Best Answer:

A shadow length   height

$\dfrac{36 \text{ feet}}{3 \text{ feet}} = \dfrac{x}{8 \text{ feet}}$

x is the height of the tree.

B Use cross multiplication to solve the proportion.

$36\,(8) = 3x$

$\dfrac{288}{3} = \dfrac{3x}{3}$

$x = 96$

Chapter 16 Review

You can practice writing some answers here. When you are done, think about if your answer is a good, better, or best answer for the question being asked. Review your answers with your teacher or tutor.

1. Gordon has cut two circular sign boards, each with a diameter of 4 feet, from a 4 ft by 8 ft sheet of plywood.

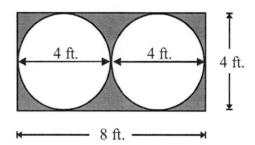

   He wants to find the area of the scrap materials represented by the shaded area above.

   A   Write an explanation that can be used to find the area, in square feet, of the scrap material.

   B   Find the area of the scrap material.

2. Kent has a large tree in his yard that he wants to cut down. He needs to know how tall the tree is to see if there is room for it to fall down without hitting his house. He also has a small tree 4 feet high in his yard that cast a 7 foot shadow at the same time the shadow from the large tree is 112 feet long.

   A   Write a proportion that Kent can use to find the height of the large tree.

   B   Now solve the proportion you wrote. Show your work.

Index

Absolute Value, 1
Acknowledgements, ii
Algebra
  combining like terms, 108
  removing parentheses, 108
  vocabulary, 94
  word problems, 96
    setting up, 100
Angles, 123
  acute, right, obtuse, straight, 124
  adjacent, 125
  complementary and supplementary, 127
  vertical, 126
Area, 142
  composite figures, 145
  squares and rectangles, 142
  trapezoids and parallelograms, 144
  triangles, 143

Base, 94

Circle, 157
  area, 160
  center, 157
  chord, 157
  circumference, 157
  diameter, 157
  radius, 157
Coefficient, 94
  leading, 94
Cone, 168
Constant, 94
Cross Sections, 169
Cube, 168
Cubes, 171
  surface area, 176
  volume, 171
Cylinder, 168

Decimals, 15

adding, 29
changing decimals to fractions, 17
changing to percents
  changing percents to decimals, 47
comparing, 15
division, 41
division by decimals, 42
multiplication, 39
subtracting, 30
Degree, 94
Dependent Events, 201
Discount, 54
  finding the sale price, 55

Equations, 105
  two-step, 105

Fractions, 14, 25
  adding and subtracting, 26
  changing fractions to decimals, 16
  changing percents to fractions
    changing fractions to percents, 48
  comparing, 14
  dividing, 38
  multiplying and dividing, 37
  simplifying improper fractions, 25
  subtracting mixed numbers from whole
    numbers, 27
  subtracting mixed numbers with borrow-
    ing, 28
  word problems, 32

Independent Events, 201
Inequalities, 111
  addition and subtraction, 113
  graphing, 111
  multiplication and division, 114
  two-step problems, 115
Inequality
  definition, 94

Integers, 1
  adding, 4
    rules for opposite signs, 6
    rules for same signs, 5
  dividing, 8
  multiplying, 8
  rules for multiplying and dividing, 8
  subtracting, 7
Interest, Simple
  $I = PRT$, 58

Like Terms (Algebra), 108
  removing parenthese, 108
Linear and Proportional Relationships, 78

Mean, 188
  finding missing data, 189
Measures of Central Tendency, 192
Median, 190
Mixed Numbers, 27
  subtracting mixed numbers with borrowing, 28
Mode, 191

Number Lines, 19
  graphing rational numbers, 19

Opposites, 2

Percents, 47
  changing percents to decimals
    changing decimals to percents, 47
  changing percents to fractions
    changing fractions to percents, 48
  decrease or increase, 51
  finding percent of the total, 50
  finding the amount of discount, 54
  finding the discounted sales price, 54
  sales tax, 57
  tips and commissions, 53
Polygons, 134
Preface, ix
Probability, 199
Proportions, 71
Pyramid, 168

Quadrilaterals, 132
  properties, 133
  types of, 132

Range, 187
Rate, 65
Rational Numbers, 13, 21
  comparing, 21
  positive and negative decimals, 13
  positive and negative fractions, 13
  positive and negative integers, 13
  whole numbers, 13
  zero, 13
Ratios, 67
Rectangular Prism, 168
Rectangular Prisms, 171
  surface area, 176
  volume, 171

Sales Tax, 57
Sample, 195
  representative, 195
Sampling
  representative, 196
  simple random, 195
Scale Drawings, 148
Scale Factor, 150
Sentence, 94
Simulation, 204
Slope, 82
Solid Figures, 168
Sphere, 168
Statistics, 187
  comparing two sets of data, 193
  mean, 188
  measures of central tendency, 192
  median, 190
  mode, 191
  range, 187
Substitution
  numbers for variables, 95
Surface Area, 176
  composite figures, 180
  cubes and rectangular prisms, 176

other prisms, 178
pyramid, 179

Table of Contents, viii
Term, 94
Tips and Commissions, 53
Tree Diagram, 206
Triangles, 131
  types of, 131

Unit Rate, 82

Variable, 94
Volume, 171

cubes and rectangular prisms, 171
other prisms, 174
pyramids
  $V = \dfrac{1}{3}Bh$, 175

Whole Numbers, 1
Word Problems
  algebra, 96
    setting up, 100
  changing to algebraic equations, 101
  rate, 65
  ratios and proportions, 72

American Book Company
THE STANDARDS EXPERTS

7 Math
Common Core

Dear Student,

Welcome to American Book Company's Common Core series! This series has been made just for you! Every book will give you practice for 100 percent of the common core standards. We have tried to explain each standard as clearly and simply as possible.

Our writers have also included Depth of Knowledge (DOK) levels in these books. You will see the levels in the chapters and the practice questions. These leveled questions will increase your ability to understand new concepts.

Also, with each grade 3–12 series, we have included free online testing from now until August 2014. If you have one of these books, the link and code are listed at the bottom of this page.

Using the code, go online and take the pretest. This is a tool you can use to find out which skills you already know. You can also determine which skills will need more work. When you finish the test, print out the results. Then give them to your teacher.

After you are done, you can use the book in different ways. You can work through all of the material in the book, paying close attention to the areas where you made mistakes. Or, if you do not have much time, you can only work on the parts where you made the most mistakes.

The online post test will also help you just like the pretest does. But, you should only take it after you finish the practices in the book.

If you are using the K–2 series, enjoy! Each book has many activities made just for you! There are also a lot of pictures and extra space for writing in your answers.

We look forward to hearing of your success as you master the Common Core standards!

Sincerely,

Frank J Pintozzi

Dr. Frank Pintozzi
Executive Vice President
American Book Company
www.americanbookcompany.com
888-264-5877

Scan this QR code with your smart device to jump to the online testing page.

To access your pretest and post test, **visit americanbookcompany.com/online-testing/test.php** and select "**7th Grade Math CCS Pretest**" or "**Post Test.**" Enter the password "**fraction**" to access these tests.